Biological control of microbial plant pathogens

Biological control of microbial plant pathogens

R. CAMPBELL
Department of Botany, University of Bristol

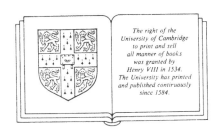

The right of the
University of Cambridge
to print and sell
all manner of books
was granted by
Henry VIII in 1534.
The University has printed
and published continuously
since 1584.

CAMBRIDGE UNIVERSITY PRESS
Cambridge
New York Port Chester
Melbourne Sydney

Published by the Press Syndicate of the University of Cambridge
The Pitt Building, Trumpington Street, Cambridge CB2 1RP
40 West 20th Street, New York NY 10011, USA
10 Stamford Road, Oakleigh, Melbourne 3166, Australia

First published 1989

British Library cataloguing in publication data

Campbell, R.
Biological control of microbial plant
pathogens.
I. Plants. Diseases. Biological control
I. Title
632'.96

Library of Congress cataloguing in publication data

Campbell, R. (Richard)
Biological control of microbial plant pathogens / R. Campbell.
    p.   cm.
Bibliography: p.
Includes indexes.
ISBN 0 521 34088 8.   ISBN 0 521 34900 1 (paperback)
1. Phytopathogenic microorganisms – Biological control.   I. Title.
SB732.6.C33 1989
632'.3—dc19   88–25812 CIP

ISBN 0 521 34088 8 hard covers
ISBN 0 521 34900 1 paperback

Transferred to digital printing 2003

PN

Contents

Contents

Preface

This book is a response to the increasing interest in biological control of plant diseases which is being shown by academic, commercial and agricultural organizations, and individuals, all over the world. The subject now receives more than a brief mention in most courses taught on plant pathology and is clearly going to be more important in the future. The excellent books by R. J. Cook and K. F. Baker will remain the standard references and there are now many review articles on various aspects in specialist journals. The present text attempts to provide an account of biological control of plant diseases that will be suitable for undergraduate students at college or university who will be meeting the subject for the first time. It is hoped that teachers at other levels will find it useful and that it will help research workers in many fields to enter the literature on disease control through biological means. The two introductory chapters attempt to set out general principles of microbial, host and pathogen interactions, and the historical and commercial background to biological control. The glossary is not comprehensive, but is designed to help those with a limited background in plant pathology and ecology.

There is a great deal of information produced for and by the agricultural industry on the chemical means of controlling diseases, as well as a vast research and teaching literature. Hopefully this book will provide some readily available information and examples of biocontrol. I by no means disparage the enormous benefits that have resulted from the correct use of pesticides, especially fungicides so far as we are concerned, but there is now a need to present a balanced argument for and against different methods of disease control. There is, rightly or wrongly, a growing public anxiety over 'pesticides' and other 'agro-chemicals', and extreme views have been expressed on both sides that have not helped the debate. Throughout this book a realistic view of what biological control can and will achieve in the forseeable future is attempted. Other researchers and teachers will no doubt disagree with my assessment of this or that control system, but at least the argument for or against will have been raised and brought to the attention of the students. Biological control is clearly not going to completely displace the use of pesticides in western agriculture, nor is there any reason why it should in many cases, but it will increasingly be a part of integrated control programmes for major world crops. In countries with less intensive agriculture, low-input systems, or subsistence agriculture the problems and aims are different and it is unfortunate that there is little information in the world literature on biocontrol under these conditions, though it undoubtedly occurs and is important. There is a great need, and great scope, here for cultivation systems and low-technology applications of biological control that may involve levels of manpower which are unacceptable in mechanized agricultural systems. Enhancement of the existing biocontrol inherent in multiple or mixed cropping and the use of organic amendments will be important.

If the information in the following pages can form a background for research workers, and stimulate and interest students to pursue and develop all forms of biological control programmes for the large number of plant diseases that occur in many different agricultural systems which are used to produce the world's food, then I will have achieved my aim.

I am indebted to my colleagues in the Department of Botany for many discussions and comments, and especially to the library staff under Sue Pettit for their continuous help. I have included references to as much of the work on the subject as was possible in a text of this length, and specific data are of course acknowledged, but there will be many research workers who recognize their results behind some generality or conclusion which is drawn, and I thank them for their unwitting help, and apologize to them for the lack of detailed reference to their work. Finally I thank Dr M. F. Madelin and Dr M. Lennartsson for reading the typescript and offering many suggestions for improvement, though they may not have agreed with all I have said and the content of the text remains my responsibility.

R. Campbell
Bristol
March 1988

1

Introduction to plant pathology and microbial ecology

1.1   Introduction

The study of agricultural microbiology has expanded in recent years and one of the areas of particular interest has been biological control. The aim of this book is to provide an introduction for undergraduate students, or for research workers in related subjects, to the biological control of plant pathogens, especially of agricultural crops. The subject overlaps with the study of microbial inoculants that may be commercially available (e.g. *Rhizobium*) or still in the development stage like those for some mycorrhizas. Indeed the problem with biological control is that it impinges on so many other subjects because the approach is always holistic: it tries to combine the manipulation of edaphic and microclimatic factors with crop husbandry, plant breeding and direct intervention with microbial inoculants to produce maximum plant growth and minimum disease. This practical, commercial bias with agricultural crops has become dominant, but biological control does of course originate in natural ecosystems where in general serious disease is the exception and pathogens are supposed to have co-evolved and to exist in balance with their higher plant host and with the other microorganisms in their environment.

Biological ways of controlling disease have therefore existed for as long as hosts and plant pathogens, and they have been the only way of disease limitation until the last few years when chemicals became available. Lime sulphur was introduced in 1802 and Bordeaux mixture in 1882. This very recent advent of pesticides, like many other new 'fashions', has led to a temporary over-reaction and their over-use in some situations. The movement now is to integrate pesticides into 'traditional' disease control systems and to add microbial manipulation, which our increased understanding of microbial ecology and plant pathology makes possible.

So biological control of plant diseases, in its widest sense, is any means of controlling disease or reducing the amount or the effect of pathogens that relies on biological mechanisms or organisms other than man. It includes (1) crop rotation and some tillage systems and fertilizer practices which affect microbes, (2) the direct addition of microbes antagonistic to pathogens or favourable to the plant, (3) the use of chemicals to change the microflora, (4) plant breeding, as it is known that changes in the plant genome may affect disease resistance and also the surface microflora (in the phyllosphere and rhizosphere). A more narrow approach is to restrict biological control to the artificial introduction of antagonistic microflora into the environment to control the pathogen. This is derived from the entomologist's approach to the biological control of insect pests by the introduction of predators to prey on a particular pest, but while this highlights the present direction of research in the biocontrol field it ignores the biological control mechanisms that have always existed in natural environments and which have been used in agriculture for many centuries. In practice most discussions on biological control do not include general agricultural methods and plant breeding systems, and certainly in what follows lack of space means that these will only be mentioned, though they undoubtedly contribute to disease control via the host plant. Most of the discussion centres on biological control that involves microbial interactions with the host or the pathogen to reduce the inoculum of the pathogen or the severity of the disease symptoms.

1.2   Ecological background

Organisms are almost always a part of a community of interacting populations or individuals. There are some communities, usually in extreme environments, such as thermal springs, that have very limited numbers of species. In general however there are predators and prey, parasites and pathogens of both animals and plants, vegetation that is eaten by herbivores and all the other various interactions with which ecologists construct their diagrams of intricate food webs or elaborate computer models of ecosystems. All this means that the abundance of one organism is partly controlled by other organisms and by the environmental factors. It is now clear that the same effects, which classically are studied in higher plant and animal ecology, also apply to microbial communities in both natural and agro-ecosystems.

If the numbers and activity of a micro-organism (a plant pathogen) is controlled by another member of the community (a saprotroph) we have biological control. Let us now examine this in rather more detailed ecological terms.

All microbes (like other organisms) occupy a niche. A niche is not a place, but an abstract concept of the role the organisms plays in the community: it is the sum of the physiological properties of the organism, of the environment or micro-environment and of the exploitable resources which together define what the microbe does. Only very rarely is it possible to define a niche in terms of a few factors. However, if we consider just one factor in the delimitation of a niche then the organism's response to it may be a simple increase in numbers (Fig. 1.1). Another species may interact with that resource or environmental factor in a different way so that its population peaks at a different level (Fig. 1.1) and the niches overlap. If more than one factor is considered then the niche is represented by a volume and the niches overlap in this 3-dimensional space (Fig. 1.2). Usually a niche is delimited by many factors, is multidimensional, and has been defined as a *n*-dimensional hypervolume in which the species can maintain a viable population (Hutchinson, 1957; see Begon *et al.*, 1986). These characteristics define the fundamental niche, the niche which the organism is inherently capable of utilizing, but the organism may be restricted by competition with other organisms for a limited quality or quantity of resources and thus it may come to occupy less than the hypothetically possible, fundamental niche and this is called the realized niche. It is the object of plant pathologists to discover the fundamental niches of pathogens and then to find ways of reducing this '*n*-dimensional hypervolume' to a

Fig. 1.1. Diagram of one facet of the niches of two populations: responses of the populations to an environmental gradient. (From Clapham, W. B. (1983). *Natural Ecosystems*. New York: Macmillan Publishing Co.)

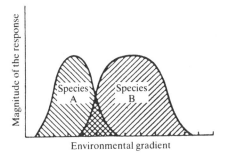

smaller realized niche by the imposition of limiting environmental factors, resources, chemical toxins and/or competing organisms that are antagonistic to the pathogen. Similarly it is important to try to understand the niches of potentially antagonistic micro-organisms so that they can be made to overlap and interact with the pathogen niche. We will return to this when we consider (section 1.3.2) competitive interactions between pathogens and potential antagonists.

The difference between the fundamental niche and the realized niche is determined by competition and predation and it is the various forms of competition that are responsible for the ability of one organism to reduce the numbers or activity of a pathogen in biological control. A pathogen whose realized niche is reduced or even eliminated is an unimportant or a dead pathogen. This is the principle of competitive exclusion: when an organism's realized niche is reduced it has been excluded from that part of the fundamental niche by the competition or predation of some other organism. If the organisms have very similar niches there will be replacement of one by the other when the new

Fig. 1.2. Diagram of the niches of two populations in respect of two different environmental gradients. One of the variables is the same as in Fig. 1.1. Note the overlap between niche volumes. (From reference as Fig. 1.1.)

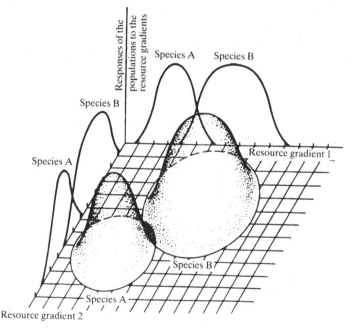

organism's fundamental niche more closely approximates to the real situation. If the niches overlap then usually one organism will prosper and the other decline, though it is possible to get deadlock situations.

Competition has many aspects but it may be important at two main stages of growth. There may be competition to establish on a fresh resource and then, once there, to acquire sufficient of that resource to survive and reproduce in the presence of other organisms. This has led to various strategies being adopted by micro-organisms (Andrews & Harris, 1986). The *r*-selected species (which are similar to the ruderal species of some authors) have a very high reproductive capacity. There are so many spores or cells that there is a good chance of them being near to any resource and they therefore compete well in dispersal (Cooke & Rayner, 1984). They are characteristic of disturbed or ephemeral sites, for example easily decomposed organic matter or root exudates where primary resource capture is important for survival. There is another group of organisms (called *K*-selected or *K*-strategists) that are characteristic of stable situations, where competition for dispersal is replaced by competition for space or for a limited resource (Begon *et al.*, 1986). Characteristically the *r*-strategists live at low population densities and as the microhabitat becomes more crowded the *K*-strategists become more important. Obviously there is rarely a completely disturbed, uncrowded situation or a completely stable, very crowded one but rather a continuum of varying degrees of *r*- and *K*-strategists (see Andrews and Harris, 1986, for further discussion of *r*- and *K*-strategists in microbial ecology). The concept of an *r–K* continuum was originally developed for studies on large animals; other systems of classification are suggested for plants. Grime (1977) suggested that ruderal species (*r*-strategists) are replaced by competitive ones (ones with a high competitive saprophytic ability) and then as the substrate becomes nutrient poor, or under unfavourable environmental conditions, these are in their turn replaced by stress tolerant species (Fig. 1.3). These categories again grade into one another and many intermediate groups have been distinguished.

Plant pathogens are spread across this whole range of strategies. There are opportunistic pathogens (ruderals) that are able to attack young or weakened plants, but are themselves poor competitors (*Botrytis, Pythium, Rhizoctonia*). There are stress tolerant pathogens: rusts living in and on leaves initially have nutrient stress and probably low water availability, and then they tolerate or control the various plant defence mechanisms. They live in or on leaves where competition is not great, because relatively few other species are so adapted to this stressed environment. Some pathogens (especially fungi) attacking living or recently felled wood are stress tolerant in these low nutrient and frequently dry environments containing plant toxins such as resins and gums. However, many of the wood rotting fungi are competitive (combative, see Cooke & Rayner, 1984) and there is a great range of strategies adopted by competitive pathogens in general. Some are specialists (*sensu* Garrett, 1970) such as *Armillaria mellea* on trees and *Gaeumannomyces graminis* on grasses and cereals, which are good at primary colonization. They do not produce a lot of spores, but they persist and defend the captured resource against secondary invaders which are often necrotrophs. Neither of these fungi are good at invading

Fig. 1.3. The possible strategies of micro-organisms in the colonization and retention of resources, showing the relationship between, and characteristics of, *r*- and *K*-selected organisms, ruderal species, stress tolerant and combative species. (Based on Grime, 1977; Cooke & Rayner, 1984; Andrews & Harris, 1986.)

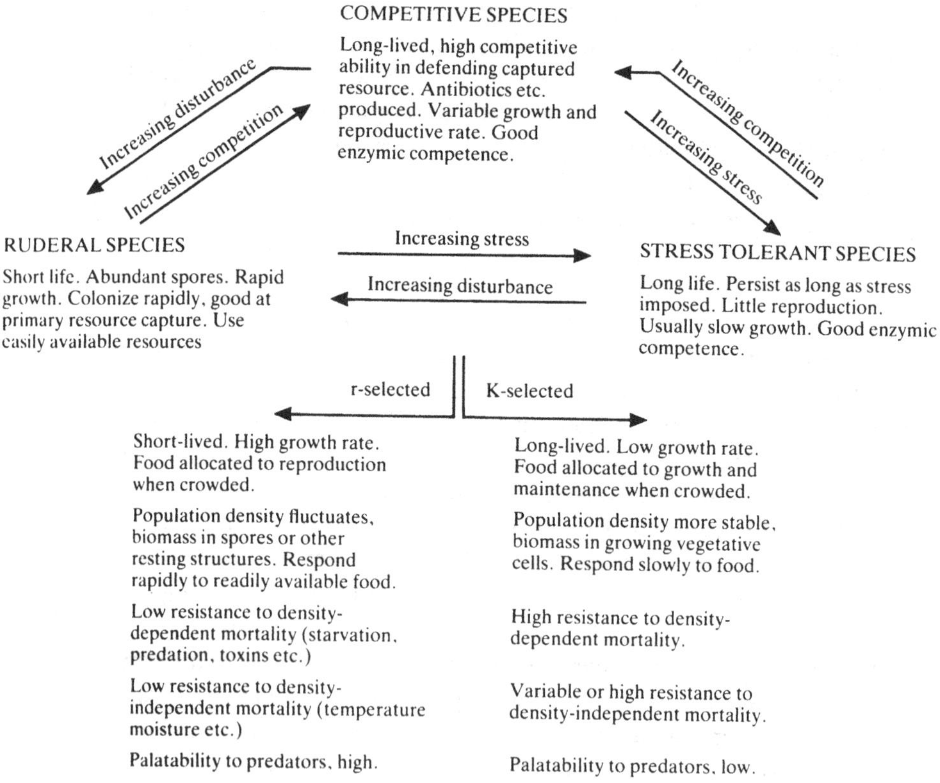

COMPETITIVE SPECIES

Long-lived, high competitive ability in defending captured resource. Antibiotics etc. produced. Variable growth and reproductive rate. Good enzymic competence.

Increasing disturbance
Increasing competition

Increasing competition
Increasing stress

RUDERAL SPECIES

Increasing stress →

← *Increasing disturbance*

Short life. Abundant spores. Rapid growth. Colonize rapidly, good at primary resource capture. Use easily available resources

STRESS TOLERANT SPECIES

Long life. Persist as long as stress imposed. Little reproduction. Usually slow growth. Good enzymic competence.

r-selected   *K*-selected

Short-lived. High growth rate. Food allocated to reproduction when crowded.

Population density fluctuates, biomass in spores or other resting structures. Respond rapidly to readily available food.

Low resistance to density-dependent mortality (starvation, predation, toxins etc.)

Low resistance to density-independent mortality (temperature moisture etc.)

Palatability to predators, high.

Long-lived. Low growth rate. Food allocated to growth and maintenance when crowded.

Population density more stable, biomass in growing vegetative cells. Respond slowly to food.

High resistance to density-dependent mortality.

Variable or high resistance to density-independent mortality.

Palatability to predators, low.

already occupied resources. Other pathogens such as the post-harvest rots (e.g. *Penicillium*) produce antibiotics that deter competitors, and yet others can infect from the soil and have a very high competitive ability (*Fusarium culmorum*). There are of course pathogens that are difficult to classify: mildews occupy the leaf, a stressed environment, yet have high growth rates and massive spore production. It is however necessary to know the strategy of the pathogen before one can seriously consider whether biological control is possible: stress tolerant and competitive species may be more difficult to control than ruderal ones or at least will require a different strategy on the part of the proposed biological control agent.

What then should be the properties of the biological control agent? In many agricultural situations there is disturbance and new primary resources are presented by removing the previous crop, by cultivation and by planting new seeds or seedlings. Even more new resources may be available if organic composts, green manures or mulches are added or retained as part of the agricultural system. A frequent need therefore is a control agent that has ruderal characteristics: for example it should (a) grow fast and have undemanding nutrient and environmental requirements, (b) be good at primary resource capture to colonize organic matter, the new plants or seedlings, (c) be adapted to disturbed, cultivated environments, and (d) have some means, usually spores or sclerotia, of surviving in the soil or on the plant near to the pathogen inoculum or the source of infection. Ruderal species are also the easiest to isolate and to culture in the laboratory, therefore they are produced by many of the protocols used by researchers whose ultimate aim is a commercial product to be grown in fermenters and back-inoculated into the field. Ruderal biocontrol agents are a good equivalent of a protectant fungicide, being in position before infection. In some situations it may be better to have a more competitive species when it is to operate against a pathogen which has already invaded the host. The pathogen then has to be displaced and the resource captured and defended, rather than colonized. Many of the potential antagonists selected by *in vitro* screening for antibiotic production (see section 2.5) will be competitive species, not necessarily good for protecting new roots or leaves. Alternatively, the control agent could reduce inoculum potential by lysis or parasitism of dormant propagules. Finally, a biocontrol agent may have to be stress tolerant, especially for use on leaves or in dry climates where soil moisture deficits may be great. Such organisms may not grow very fast or colonize well, so the inoculum may have to be applied in massive doses to give cover without depending on growth or movement of the organism in the environment. Examples of these different sorts of control agents will be given in later chapters.

The strategy of both the pathogen and the proposed control agent should be considered right at the start of a development programme. Obviously there may be several different stages in the pathogen life cycle that may be attacked (see section 1.4) and combinations of different control agents with different strategies could be used to cover situations where ruderal, competitive or stress tolerant characteristics were needed as the crop grew and developed, as the disease progressed or as the seasons advanced.

The difficulty, indeed the fallacy, of trying to compartmentalize ecological strategies in this way is well known, but if the pathogen is at least thought about in these terms it may become more obvious whether we are trying to protect the plant from primary colonization, kill the pathogen in the soil or invade an existing infection on the plant. It may then be possible to modify the isolation, screening and selection procedures so that at least there is a chance of growing the right sort of biocontrol agent or of detecting its activity in the environment. In the following sections of this chapter we will consider in more detail exactly what makes an organism good at colonizing a primary resource, and what are the ways in which one organism antagonizes another.

1.3   Mechanisms of biological control

There are many ways in which an antagonistic organism can operate (Elad, 1986): rapid colonization in advance of the pathogen or subsequent competition or combat may lead to niche exclusion, antibiotics may be produced or there may be mycoparasitism or the lysis of the pathogen. In addition some micro-organisms may act simply by making the plant grow better, so that even if the disease is not cured its symptoms are at least partly masked.

1.3.1   *Colonization and inoculum*

Colonization of a plant can only occur from inoculum either resident in the environment or brought there by wind, water, animals or man. Much work has been done on pathogens, mostly fungi, in connection with the epidemiology of plant disease (Scott and Bainbridge, 1978). Very little is known about the epidemiology of organisms introduced into environments foreign to them, though this is now receiving attention, especially with the possibility of releasing genetically en-gineered organisms.

Plant roots and shoots grow at the tip, so new more or less sterile material is continuously produced and the root or shoot is older towards the original position of the seed. There are some exceptions to this such as *Pythium* colonizing seed embryos, so that the emerging root is already infected, or *Taphrina* (a leaf-inhabiting ascomycete) over-wintering in buds and infecting leaves as they open in spring. However, in general, we are talking about the arrival of microbes onto an unpopulated surface or organ and the colonization of that surface if the nutrients, environmental conditions, etc. are suitable for that organism. This is usually the colonization of a primary resource by a ruderal species. Later arrivals will not find a vacant niche, they will find resources and space already occupied and they must be competitive.

Dispersal of micro-organisms to above ground parts of plants is mainly by airborne spores or in splash dispersal (Ingold, 1978; Horsfall & Cowling, Vol. 2, 1978; Campbell, 1983, 1985). The surface structure of the leaf is most important in determining the ease of landing. Small hairs and other projections can extend the laminar boundary layer so that only very fast moving or very large spores can penetrate: this makes the arrival of the small bacteria difficult. Alternatively very large spines may project above the boundary layer and since they have a compara-tively small diameter they will trap quite small particles. If the surface is waxy, as it frequently is, then water is repelled from the surface and any microbes in the water do not stick to the surface.

As has been mentioned previously the leaf of temperate plants is a stressed environment with low water and nutrient levels, and the microbes are therefore forced into slightly protected microhabitats, such as under spines and in epistomatal cavities (Campbell, 1985). The severe nutrient shortage means that there is competition for those that are present: this will be considered in detail in section 3.5, but Fig. 1.4(*a*) shows that the germination of some pathogenic fungal spores can be reduced by bacteria which accumulate the available amino acids. With other pathogens (Fig. 1.4(*b*)) the germination is not affected by bacteria or nutrients but the elongation of the germ tube is much reduced by competition with the same bacterium. If this bacterium was used as a biocontrol agent it could be selective in its control of different pathogens. Some pathogens have escaped, or become tolerant to, this competition by tapping a source of nutrients not available to the surface saprotrophs. They go inside the plant and extract nutrients. This may lead to very dense surface growth in a few pathogens (e.g. the powdery mildews, *Erysiphe*). Biocontrol on temperate leaves therefore requires

Fig. 1.4. The effect of *Pseudomonas* sp. on the germination of conidia. (*a*) *Cladosporium herbarum*: ■ — ■ = % germination, ▲ — ▲ = ^{14}C amino acids uptake by the bacterium released as carbon dioxide, ○ — ○ = proportion in cells only. (*b*) *C. dematium* as (*a*) plus □ — □ = length of germ tube. (From Blakeman, J. P. & Brodie, I. D. S. (1977). *Physiological Plant Pathology* **10**, 29–42.)

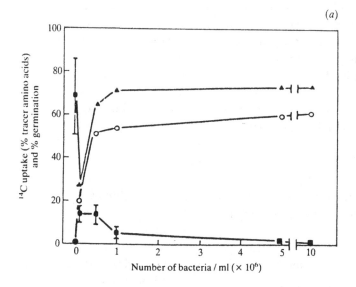

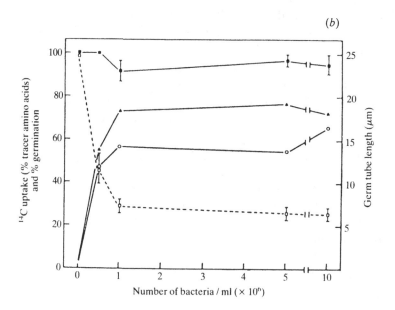

stress tolerant organisms, or if ruderals are used to get early, protective colonization then nutrients may have to be added and the water levels or atmospheric humidity kept very high.

In the moist parts of the tropics the situation is very different, for water can be in abundance and leaching from epiphytic autotrophs can raise nutrient levels. Here the main need might well be a competitive biocontrol agent to establish itself in an occupied microhabitat.

Most agents for use on the above ground parts of plants will be applied as dusts or sprays, where the drop or particle size and velocity may have to be carefully controlled if the organism is to reach the surface. These considerations will be enlarged upon in Chapter 3, but this brief mention may give a slightly more practical basis for the present theoretical discussion.

Below ground the situation for colonization is very different. The plant usually lives in fairly moist soil for at least some of the year, though there may be periods of very low water potentials. Nutrients are exuded from the roots much more than from the leaves. The stress is therefore less and the competition may be greater. On the other hand soils are often very heterogeneous, on both a micro- and a macro-scale, and on agricultural land there may be frequent disturbance as a result of cultivation. These factors may favour ruderal organisms. It is therefore not always easy to predict the dominant factor which may be controlling the type(s) of organisms present.

The natural inoculum below ground is often very abundant but is not usually brought to the root. In contrast to the airborne situation, it is the plant that grows towards the inoculum and arrival may be largely a matter of chance contact, though colonization may thereafter be active and competitive. The spores of some microbes can germinate in response to the presence of plant roots, but others may grow towards a nearby root, and some important soil fungi (*Phytophthora*, *Pythium*) have zoospores that may show positive chemotaxis. Microbes can also be moved around by animals in the soil. However, in general a soil inoculum sits and waits for the plant rather than being blown around in the air looking for one!

The concept of inoculum potential has been largely developed for soil-borne plant pathogens, though it is applicable to other sorts of pathogen as well. The problem is to attempt to define how much inoculum there is in the soil, and hence how badly a plant will be infected or how well an organism will colonize and give protection. If there is some way of measuring inoculum it may seem obvious that the more there is the worse the infection, and this is sometimes true. However, there may need to be a certain amount of inoculum to cause disease; often this is a case of overcoming the plant's defence systems only by multiple attacks at several different points or of needing several propagules at each point to establish infection. There is therefore a threshold value, below which there is no infection and above which there may be quite a lot, with not much gradation between. This is illustrated in Fig. 1.5, where below 0.01 mg of propagules per gram of soil there is no infection but thereafter the disease increases greatly. The particular value required to produce infection varies with the pathogen and with the host, depending presumably on the relative virulence of the pathogen and the degree of resistance to infection shown by the particular host. Thus in Table 1.1 *Rhizoctonia* requires a minimum of only 0.01 units per g to infect sugar beet but 0.07 to infect cotton.

Fig. 1.5. Mean number of lesions per plant in relation to inoculum concentration (weight in mg, log. scale) in fumigated soil. Inoculum particle sizes were: ● = 0.75 mm, ○ = 0.35 mm, □ = 0.2 mm, ■ = 0.128 mm, △ = 0.053 mm. (From Wilkinson, H. T. *et al.* (1985). *Phytopathology* **75**, 98–103.)

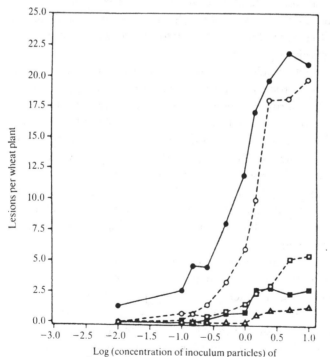

Log (concentration of inoculum particles) of
the take-all fungus (*Gaeumannomyces graminis*)

Alternatively, cotton will be diseased by only 0.01 to 0.06 units per g of *Phymatotrichum* but requires 50 to 400 units of *Verticillium albo-atrum* under some conditions. This latter point is important: the minimum amount required will rise under conditions of temperature, water availability, etc. that are unfavourable to the pathogen, and will fall if the host is stressed or partially incapacitated by some other agency.

Apart from the threshold and the effects of environment there is also a factor concerned with the vigour of the inoculum. If the propagule has been dormant in the soil for many months, or even years, it may have used most of its food reserves and not grow very quickly or produce enough enzymes to degrade host cell walls. Apart from endogenous food reserves the pathogen may be dependent on food available in a saprotrophic situation. Figure 1.5 shows the effect of inoculum particle size on the infection of wheat: larger particles contain more food so more of the pathogen survives, has a better food base to grow out through the soil to the root and is also better protected by a large

Table 1.1. *Threshold population densities that are needed to produce disease by selected soil-borne pathogens*

Pathogen	Host	Population density (units/g)
Sclerotium rolfsii	sugar beet	0.005–0.05
Phymatotrichum omnivorum	cotton	0.01–0.06
Rhizoctonia solani	sugar beet	0.01–0.09
	cotton	0.07–0.13
Gaeumannomyces graminis	wheat	0.01–0.3
		0.05–0.11
Verticillium albo-atrum	cotton	50–400
	cotton	0.03–50
	mint	10–100
V. dahliae	potato	10–130
Phytophthora cinnamomi	fir seedling	1–30
	pineapple	1–3
Plasmodiophora brassicae	cabbage	>10
Fusarium solani f. sp. *phaseoli*	bean	1000–3000
F. roseum f. sp. *cerealis* 'Culmorum'	wheat	100–3000
F. solani f. sp. *pisi*	pea	100–6000
Thielaviopsis basicola	citrus	1000–8000
Pythium ultimum	pea	100–350
	pea	100–1000

From Baker & Cook, 1974.

particle from possible desiccation. Near to the threshold, where almost nothing is able to cause infection, the size of the particle has little effect. As the total amount of inoculum (weight of fungus per g of soil) is increased then disease increases, but larger particles cause much more disease than the smaller ones. Small particles also have a higher threshold.

The amount of disease is therefore determined not only by the actual quantity of inoculum but also by many other factors that have been combined in the concept of inoculum potential (Garrett, 1970; Horsfall & Cowling, Vol. 2, 1978). Inoculum potential is the sum of all the factors that contribute to the energy available for infection of the host by the pathogen and it is one of the main determinants of the amount of disease produced. As such it has been much studied, both empirically by field pathologists and also in attempts to use it in computer models to study and predict the course of disease epidemics and invasion of plant tissues by pathogens (R. Baker, in Horsfall & Cowling, Vol. 2, 1978).

Once the pathogen has reached the root, the colonization is likely to be affected by competition with microbes already in residence (see below).

Colonization by micro-organisms artificially introduced into the environment, such as biocontrol agents, may be very different. Firstly, the inoculum potential can be affected by the amount of the inoculum supplied, by controlling the growth conditions, formulation and storage to give good viability and by ensuring germination under conditions that will be advantageous to the antagonist. Secondly, the inoculum can be placed so as to be in a favourable position for colonization. It may be sprayed on the soil or mixed in as a pellet or granule to control pathogens in the soil, allowing natural contact or spread to take it to the microsite in which the pathogen lives. Alternatively, the inoculum may be put on the seed, as a pellet, powder coat or liquid, so that the emerging root or shoot meets it as it grows. In the case of aerial plant parts the inoculum is usually applied as a surface spray often in very high amounts so that colonization in this difficult habitat is not really necessary, the organism being everywhere from the start.

In general the aim of an artificial inoculum application is to use an *r*-strategist with good competitive colonization, placing a given minimum necessary amount as close as possible to the proposed point of action, in a form in which it will survive for the required length of time (section 5.1). This latter point is not to be confused with the 'shelf life' – the storage time in a dormant state between manufacture and use.

1.3.2   *Competition*

Competition occurs when two (or more) organisms require the same thing and the use of this by one reduces the amount available to the other. Thus micro-organisms may compete for nutrients: one organism (because of better uptake mechanisms or better extracellular enzymes) gets most of the nutrients and grows, while another has insufficient and dies. This is known for both carbon and nitrogen sources. Competition is also possible for oxygen, space and, in the case of autotrophs, light. An essential point of the definition is the deprivation of one of the organisms: if there are excess nutrients so that all have enough there is no competition. Micro-organisms cannot therefore compete for water; they may need water but they do not really affect the amount present or the distribution of it as a large higher plant would. As far as a micro-organism is concerned, water is either present in sufficient quantity or not and neither the micro-organism nor its potential competitors can affect that quantity. Microbes may however compete for space in which water levels are suitable or optimum.

Thus on temperate leaf surfaces very little of the space is occupied (<1% usually) because much of the surface has intolerably low water or nutrient levels and the competition is not for total space, but for the space which is utilizable. On roots the limiting factor is again nutrients; any one root produces a given amount of exudate (section 5.1) which allows the growth of a given biomass and this is attained regardless of the initial inoculum levels (Fig. 1.6). If more than one organism is growing on the root then their total biomass will be the same as one growing alone if they use similar nutrients (Fig. 1.7). One of the best examples of space competition on roots is that of ectomycorrhizae (section 5.4) on some tree roots where the fungus sheath effectively covers the entire root so that the rhizosphere is occupied and any other organisms must colonize the fungus surface rather than the root *per se*; the mycorrhizal root has its own sphere of influence which is different from the non-symbiotic root. Alternatively the presence of bacteria can affect the amount of colonization by mycorrhizal fungi, though this is probably antibiosis rather than competition for space.

There are other examples of ectotrophic growth where some fungi, especially some root pathogens, use it to avoid nutrient competition and the host reaction to infection. There is a group of fungi belonging to the genus *Gaeumannomyces*, its imperfect (anamorph) state and related deuteromycete fungi (*Phialophora*) which grow very rapidly on the outside of roots, sending down feeding hyphae into the root cortex at

Fig. 1.6. Effect of three levels of inoculum on the growth rate and final population size of a fluorescent *Pseudomonas* on barley roots. (From Bennett, R. A. & Lynch, J. M. (1981). *Current Microbiology* **6**, 137–8.)

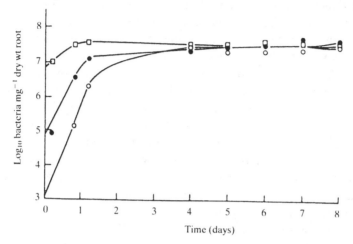

Fig. 1.7. Numbers of bacteria per g dry root for bacteria inoculated onto gnotobiotic plants. ○, ● = *Serratia marcescens*; □, ■ = *Flavobacterium* sp. Open symbols are for bacteria inoculated onto separate plants, closed symbols are for both bacteria on the same plant. Vertical bar is LSD, P = 0.05. (From Turner, S. M. & Newman, E. I. (1984). *Journal of General Microbiology* **130**, 505–12.) Permission from the Society for General Microbiology is gratefully acknowledged.

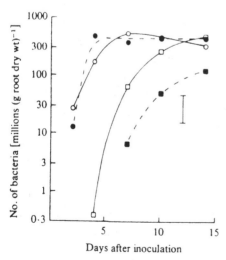

intervals. By the time the host reacts, the hypha has grown away from the region, using a food supply inside the plant that is not available to potential competitors on the root surface. The hyphae of these fungi tend to grow rather more up the root towards the crown of the plant rather than down towards the root tip. Most hyphae grow much more slowly than these and in general fungi and bacteria cannot maintain a constantly favourable food supply from the exudates by growing with the root tip – it is still competitive colonization rather than competition between existing organisms.

There is competition for space in the later stages of decay of resource units such as large tree trunks and woody branches and roots (Cooke & Rayner, 1984). Initially heartwood is low in readily available carbon, nitrogen and phosphorus though sapwood contains more of these essential elements. There are also various toxins such as tannins, gums and resins that may inhibit growth. This is thus a highly stressed environment and the colonizers are stress tolerant with specialized enzyme systems. Some may be very specific and may avoid competition by colonizing particularly difficult species. Thus *Fistulina* attacks only oak, which is a very hard wood high in tannins. After this phase there may be a more general invasion by competitive species which can form complex patterns of decay columns within the wood and they may out-compete other organisms for the space in the log or they may actually displace existing colonists. Some species may be 'equally matched' in this fight and then a stalemate may result with the interaction deadlocked, neither competitor gaining advantage. Such competition may occur between fungi of different genera, between different species and also between genetically different strains of one species. The different mycelia are often separated by dark coloured zone lines of specialized mycelium. As we shall see in Chapter 4, the stem, both woody ones such as we have just been discussing and herbaceous ones, is one of the places where there are successful biological control systems. Most of these are not competitive biocontrol agents, but are ruderals which must be applied before infection into the vacant niche of an unoccupied stem. There are no biocontrol agents that have a high enough competitive ability to displace a pathogen which is in possession of this resource.

Competition for oxygen is another possible mode of interaction. It is rather difficult to measure on the microhabitat scale but it is known to occur in some germinating seeds. Cereal seeds sown into high organic matter soils or where straw is decomposing are anyway short of oxygen and this makes them especially liable to leak nutrients, usually from the micropyle. This region is then colonized by an assortment of organisms but especially the fungus *Gliocladium* which grows on the exudates and makes the oxygen deficiency worse. There are various microbial inoculants and chemical treatments that have been used to combat this. Oxygen competition may also account for some of the disease control obtained from an increase in general microbial activity when organic matter is added to soil, though there are other mechanisms also operating (section 5.3).

The discussion up to now has been mainly concerned with competition with the pathogen on or near the host, but some pathogens also have a saprotrophic growth stage in soil or on plant debris. Two separate things should be distinguished here, firstly the saprotrophic, dormant survival of a propagule (Garrett, 1970) such as spores or sclerotia within the remains of the host. *Sclerotium rolfsii*, for example, survives as sclerotia in plant debris, but these sclerotia may be specifically attacked by the biocontrol agent *Sporidesmium sclerotivorum* (section 5.6.1) and other fungi such as *Trichoderma*. Similarly *Gaeumannomyces graminis* survives as dormant mycelium in the stem bases of wheat plants, which were attacked the previous growing season. Secondly there may be active saprotrophic growth of the pathogen in competition with the normal saprotrophic soil population, which is exploiting the same resource(s). This second group of pathogens is exemplified by *Rhizoctonia*, which actively grows in the soil as a saprotroph, but may also be a serious pathogen: this seems to be a successful strategy, because *Rhizoctonia* is very widespread and causes a lot of diseases.

Some pathogens may adopt a combination of both these strategies. They may keep a food base but at the same time grow out into the soil from that base and use it to increase their inoculum potential or competitiveness in the colonization of new resources. Thus *Armillaria mellea* lives in dead trees and grows through the soil as rhizomorphs that are more or less independent of food availability in that soil: the rhizomorphs may attack and colonize a tree to form a new food base. In this manner the organism keeps the security of a food base but at the same time finds new food without the risk of starvation if the search is not immediately successful. *Fusarium nivale* lives in wheat stem bases and has a similar strategy, though it is less specialized and does not produce rhizomorphs.

A particular form of nutrient competition involving iron has been proposed as a mechanism of biological control. There can be competition for ferric iron by the production of special iron-chelating compounds called siderophores in iron-limited environments, such as arable soils which are on limestone rocks or which are limed to improve the aggregate structure in clay soils: they have a high enough pH to precipitate most of the ferric iron as hydroxide. Siderophores are produced by many organisms, including all sorts of micro-organisms and higher plants, to assist in the uptake of iron. Different siderophores differ in their affinity for iron (and other cations) so there can be competition between siderophores and those with the highest affinity will sequester all or most of the iron. If an antagonist can produce a better siderophore than the pathogen then the latter could be deprived of iron, and therefore grow less well. There is the possibility of also depriving the plant of iron but maybe the plant has a better siderophore than any of the micro-organisms. There are complications to this theory in that organisms can sometimes use each other's siderophores because their membrane transport systems are compatible. The siderophore effect may also act through micro-organisms other than the pathogen and its antagonist (see examples in sections 5.8 and 6.3). There also seems to be interactions with the host's defence systems. The activity of siderophores is linked to more general nutrient competition in some cases. Firstly, siderophore production itself seems to require quite a high level of nutrients, so it may be restricted in carbon limited situations, such as soil. The germination of *Fusarium* chlamydospores requires exogenous nutrients and is inhibited by *Pseudomonas* species that produce siderophores. Mutant strains, which do not produce siderophores, may still give some germination inhibition by nutrient competition. However, despite these complications, there are clear cases where antagonists work, but mutants without the ability to produce siderophores do not.

Saprotrophic survival and competition have been talked about in terms of interactions between organisms, but it should be remembered that the environment also plays a considerable part in the outcome of these events. The availability of iron has been discussed above and other nutrients, especially the type of nitrogen and its availability, have been much studied in relation to saprotrophic survival and competition (Garrett, 1970).

From these initial considerations of the effects of competition it is clear that there are many different possibilities for biological control such as (1) reducing inoculum potential either by lysis (see below) or by nutrient competition (e.g. fungistasis, see below), (2) increasing saprotrophic competition for initial resources in substrate colonization and (3) reducing the actual amount of the pathogen in either the dormant survival or pathogenic growth phases. It is necessary to know which of these are important in a particular pathogen so that, for example, you know where to place the inoculum of your antagonist in relation to the host and what sort of antagonist you should be looking for (ruderal, stress tolerant or competitive). The most common actual means of competition, apart from more efficient nutrient uptake or oxygen acquisition etc., are antibiosis and lysis. Successful competition may mean the possession of some means of antagonizing the competitor by poisoning or killing it, and these mechanisms must now be examined in more detail.

1.3.3   *Antibiotics and endolysis*

Lysis is the complete or partial destruction of a cell by enzymes. It has been much studied in relation to the destruction of invading organisms by defence mechanisms in the blood of animals, but for our purposes we may distinguish two types, endolysis, and exolysis. Endolysis (also called autolysis) is the breakdown of the cytoplasm of a cell by the cell's own enzymes following death, which may be caused by nutrient starvation or by antibiotics or other toxins. Endolysis does not usually involve the destruction of the cell wall (Fig. 1.8). Secondly, there is exolysis (also called heterolysis) which is the destruction of a cell by the enzymes of another organism. Typically exolysis is the destruction of the walls of an organism by chitinases, cellulases, etc. and this frequently results in the death of the attacked cell. In exolysis the death is caused by the lysis, but in endolysis the death is the cause of the cell's own lysis. There can be some overlap between the terms when a bacterium colonizing a hypha, for example, produces an antibiotic that causes endolysis and at the same time produces a chitinase that destroys the fungal wall so that both forms of lysis occur at the same time, and it may be difficult to determine exactly what is happening. We will discuss endolysis now and exolysis in the next section (1.3.4).

Endolysis may be caused by 'normal' death from old age or the use of all nutrients in that part of a resource. It may be caused or hastened by nutrient competition from other organisms as discussed above. It may also be caused by an untimely death brought about by toxins from another organism. These toxins are often antibiotics (which operate at low concentrations: less than 10 ppm) and should be distinguished from such things as production of hydrogen ions to change pH or the production of ethanol, which is required in comparatively high concentrations to be toxic or to inhibit growth (Baker & Cook, 1974).

There are volatile antibiotics and also gaseous products like ethene (ethylene) and hydrogen cyanide (section 5.8) that affect microbial growth, which are active at low concentrations but are not generally considered as antibiotics. True antibiotics are perhaps the most studied mechanisms of antagonism between micro-organisms. They have assumed great importance in medicine for the control of animal (including human) diseases and, therefore, enormous investments have been made by pharmaceutical companies. There is very extensive literature listing and describing the commercial production, medical use, biochemistry and so on of all the many thousands of antibiotics that have been discovered as the result of the extensive searches for these microbial products. Many of the organisms which produce commercially valuable antibiotics were initially isolated from soil, though the present industrial strains may be far removed from the original. It is possible to isolate antibiotic-producing organisms from leaves and other plant parts, but they are most common in soil. It seems obvious, therefore,

Fig. 1.8. Lysis (probably endolysis) of hyphae of the root pathogen *Gaeumannomyces graminis* in the presence of bacteria which produce antibiotics in culture. Both organisms are growing on the root surface (bottom) within the mucilage layer, the boundary of which is shown with attached clay particles.

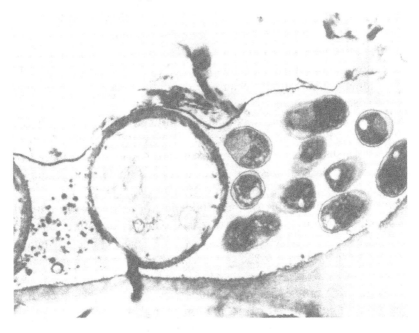

that antibiotics are important in the environment, but this is far from proven. Indeed there are very few examples known where the production of antibiotics has been demonstrated in natural environments, even though up to 52% of soil isolates may produce inhibitory substances when first isolated (Gottlieb, 1976; Williams and Vickers, 1986).

There are many possible explanations of this paradox. Firstly antibiotics are produced most abundantly in rich media, especially if the different nutrients are not in balance for normal growth. A period of rapid growth followed by starvation may be needed. The soil, and most other microhabitats associated with plants, is carbon limited: microorganisms are dormant in many natural environments because of carbon or nitrogen limitation (Campbell, 1985) and may not produce antibiotics. Soils amended with organic matter or other readily available carbon sources may produce detectable amounts of antibiotics, while root exudates and the concentration of materials by adsorption at surfaces may also allow sufficient nutrients for their production.

Even if produced in natural environments antibiotics may be adsorbed on to clay or organic colloids, which may concentrate small amounts to locally effective levels but prevent their detection in the bulk soil. This may be reflected in the fact that the most dramatic effects of antibiotics, in culture anyway, are usually demonstrated in rather poor media which have less organic materials to adsorb the antibiotic. There is a conflict between the production of antibiotics in initially rich media and their effectiveness in situations with few possible adsorption sites. Such conditions of growth and starvation with variable adsorption could be met in natural situations because of the heterogeneity and the small scale of spacial variations.

It is also possible that microbial antibiotics produced are rapidly broken down by enzymes in the soil. However, if cell-free culture filtrates from laboratory grown microbes are added to soil there are sometimes antibiotic effects so it seems that they should operate and be detectable if they are really there in reasonable amounts.

An argument in favour of the production of antibiotics in natural environments is that it is unlikely that the genetic information for their production would survive if it was not of some considerable advantage to the organisms. Therefore the very existence of antibiotics implies that they are used.

Despite these problems, and an embarrassing lack of evidence, it is generally assumed that antibiotics exist in natural environments and are active there. There are patents on micro-organisms and their antibiotics for use as biocontrol agents (see section 5.6.3). A very popular way of screening for potential control agents is to look *in vitro* for antibiotic inhibition zones produced during growth on agar media (see Chapter 2).

The antibiotics that are detected *in vitro* are very diverse and may be specific for particular target organisms or more general with a very wide spectrum of activity. Some target organisms are known to be comparatively little affected by antibiotics, for example *Fusarium*. *Pythium* is more sensitive to antibiotics produced by fungi than to those produced from bacteria. Clearly this sort of information is of use in directing the search for control agents that may work by the production of antibiotics.

The observable effects of antibiotics in culture are as varied as their origins and their chemical nature. There is generally a reduction or cessation of growth or sporulation, or a reduction in germination. This may be accompanied by various distortions of the hyphae of an affected fungus, changes in branching patterns of colonies, the production of specialized growth forms such as pseudoparenchymatous tissues and the deposition of assorted by-products from the affected metabolism. If the antibiotic causes death then endolysis of cells may also occur. Examples of the effects that antibiotics produce in culture are shown in Fig. 1.9. Notice that the effects of antibiotics may occur at some distance from the organism producing them, so this can be competition at long range.

Fig. 1.9. *In vitro* assay for antibiotic production. Different strains of the pathogen have been inoculated at right and left and a potential antagonist is inoculated at the top and bottom. There is a wide inhibition zone caused by the diffusion of antibiotic(s) from the antagonist colony.

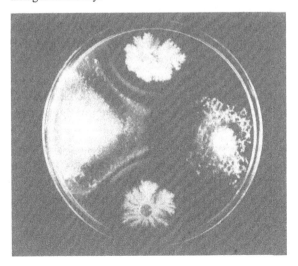

In contrast most nutrient competition (see above) and parasitism (see below) requires very close proximity if not actual contact.

Finally it should be remembered that antibiotics are not the sole prerogative of selected antagonists. It has already been said that many soil organisms produce them, so an introduced antagonist may itself be antagonized and made ineffective. Some pathogens are known to produce antibiotics that can restrict the activity of potential antagonists or competitors. *Cephalosporium gramineum* occupies the bases of wheat stems and excludes other organisms, especially from the vascular bundles, by the production of antibiotics, and artificially produced mutants of the pathogen which are deficient in antibiotic production are not able to retain possession of this food base (Baker & Cook, 1974).

1.3.4   *Mycoparasitism and exolysis*

Antagonists may operate by simply using the pathogen as a food source: if the pathogen is a fungus then the antagonist is a mycoparasite and usually possesses chitinase to break down the walls of its host. If the pathogen is an oomycete (e.g. *Pythium* or *Phytophthora*) then cellulase(s) are needed.

Perhaps the best known mycoparasite is the fungus *Trichoderma* which has been suggested as a biocontrol agent against many soil pathogens (Chet & Henis, 1985: in Parker *et al.*) and is one of the few agents at present (1987) commercially available. The hyphae of *Trichoderma* may penetrate resting structures such as sclerotia or may parasitize growing hyphae. In the latter case the hyphae grow alongside the host and send out side branches that coil around the host hypha. Penetration of the wall has been shown in some cases and assumed in others (Fig. 1.10). Other soil fungi can coil round hyphae of pathogens and produce death of the latter, sometimes without obvious evidence of holes in the attacked hypha (Fig. 1.10).

Darluca filum, *Tuberculina maxima* and *Verticillium lacanii* attack many species of leaf pathogens especially some of the rusts (Cullen & Andrews, 1984, in Kosuge and Nester, see section 3.6.3) and prevent sporulation.

Amoebae that live in soil (especially *Arachnula*, *Acanthamoeba* and vampyrellid amoebae) may parasitize hyphae and spores by cutting holes in their walls and sucking out the cell contents (section 5.7). While considering protozoa, there are also mycophagous ciliates known (of the genus *Grossglockneria*), though their potential for biocontrol has not been tested.

There are some doubts on the use of organisms causing exolysis.

Fig. 1.10. (*a*) Hyphae of *Arthrobotrys* coiling around a hypha of *Rhizoctonia* that has died and collapsed. (Photograph by Norkrans-Hertz, B., Persson, Y., University of Lund, Sweden, and Campbell, R.) (*b*) Hypha of *Sclerotium rolfsii* that has been parasitized by *Trichoderma*. The hypha of the latter has been removed to reveal the penetration sites. (Photograph courtesy of Chet, I. From Elad, Y. *et al.* (1983). *Phytopathology* **73**, 85–8.) (*c*) and (*d*) Degradation of the hyphal wall of *G. graminis* by bacteria. The hypha running from left to right has been colonized by bacteria (*c*) and only a few fragments of the wall are left (*d*).

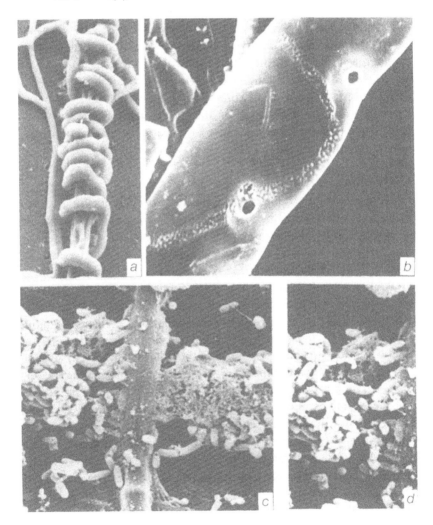

Firstly some mycoparasites produce antibiotics as well as causing holes (e.g. *Gliocladium virens*) and there are reports of these fungi that have been mutated experimentally to remove their capacity to form coils, yet the biocontrol still works. This must cast doubt on the importance of mycoparasitism in the overall control by this fungus. Secondly, as far as is known, they require contact or very close proximity for the production or the induction of the necessary enzymes, and therefore a considerable biomass of the host must be present for a period of time. This means that they are not thought to be very useful for fast growing or transitory pathogens which may evade parasitism by growing away or getting inside a leaf or otherwise hiding where the amoebae or fungus cannot reach. However, they can be used against resting structures such as spores, sclerotia and hyphae surviving in dead host tissues and therefore present for some time, often in considerable quantities. Parasites may also be useful to invade existing pathogen lesions, not to control the present infection but to reduce spore production and so reduce inoculum for the next infection.

1.3.5   *Fungistasis*

Fungistasis is the imposition of dormancy, especially for fungal spores, by nutrient limitation (Lockwood, 1986). Most commonly this involves a shortage of available carbon. Many pathogens produce resting structures of various kinds that remain dormant in the soil until nutrients are available. The saprotrophic microflora may reduce available carbon levels and impose fungistasis on the pathogen, preventing its germination and subsequent infection. One of the best examples of this is in soils where the competition for carbon amongst *Fusarium* species leads to a reduction in the disease (section 5.2.1).

The corollary to fungistasis is that the addition of readily available carbon to the soil can permit the germination of dormant spores. The key word here is *available* carbon. The addition of organic matter (as a green manure, compost or natural litter) may so stimulate microbial activity that intense competition may develop, leading to carbon limitation and fungistasis. It is only when available carbon is added above the needs of the saprotrophic competitors, that germination of pathogens may be stimulated as fungistasis is broken. The practical use of fungistasis is therefore in the manipulation of the carbon status to encourage the saprotrophs, but not the pathogens.

This completes the brief discussion of the main modes of antagonism which many examples in later chapters will examine in more detail.

1.4   Pathogen life cycles

The life cycle of pathogens has been divided into many stages, but basically there are four divisions (Andrews, 1984: Fig. 1.11). Infection of the host involves the germination of the spore or other propagule after it has found and possibly recognized a new host. These processes are affected by weather conditions, by the nutrient status of the host surface at the infection court, and by the recognition systems and initial defence reactions of the host as described below. Secondly, there is a feeding and growth stage of the pathogen during which the host may cause further inhibition. Thirdly, there is a stage of spread, which for fungi may be by asexual spores, often air or water borne, which are particularly affected by the weather. Viruses also have enormous reproductive potential, at least as far as numbers are concerned, for the infected host may produce up to 5000 virus particles per cell per hour which far exceeds any reproductive rate of fungi or bacteria. However the virus may require a vector for its dispersal rather than being independent like the fungi and most bacteria. Lastly, there is a survival stage which may be combined with the previous reproductive stage, or may be a new type of spore (e.g. a sexual resting spore) or a specialized structure such as a sclerotium. Obviously not all pathogens have all these possible stages in their life cycle. Conversely there are those which may have multiple spore stages on different hosts, such as some rusts.

Different sorts of pathogen, and different stages of the life cycle of any one, can be controlled to different degrees and by different agents. Thus the resting stages of a fungus, which may be dormant spores with thick melanized walls, are not subject to control by competition for nutrients but, as outlined above, they may be parasitized. However, at or soon after germination the spores are very sensitive to nutrient conditions and may suffer from competition. The mycelium is also vulnerable, but an obligate biotroph may avoid this by entering the plant and escaping from saprotrophs growing on the outside: only the biotroph can exist within the plant without seriously triggering the host defence mechanisms. In the case of a rust, there may therefore be a very short period when there is any chance of attacking it with a biocontrol agent – just that time between germination and penetration. The application of a biocontrol agent would have to be very carefully timed or the organism would have to respond to the same germination triggers as the pathogen. Some of the rusts present a further problem in that they have different spore stages on different hosts at the same or different times of the year. This may require more than one control

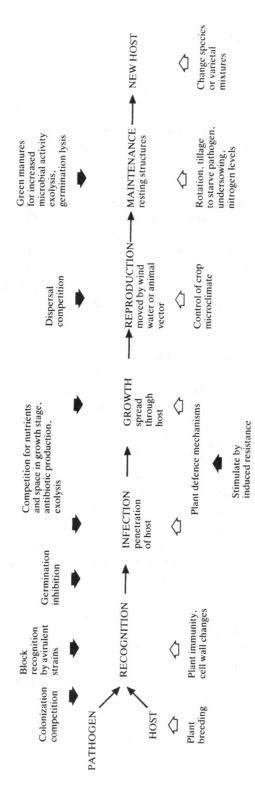

Fig. 1.11. Pathogen life strategies and possible means of host defence. Stages in pathogen life cycle: ⬦ points of control by plant host or agricultural practice: ◆ point of control by biological means. (Based on Andrews, J. H. (1984). *Advances in Plant Pathology* **2**, 105–30.)

agent, but in general rusts are difficult to control biologically because they are so well adapted to their hosts (see section 3.2).

Necrotrophic pathogens are an easier proposition for they usually spend most of their life cycle where saprotrophs, which is what most biocontrol agents are, can get at them. Even if they are leaf pathogens they frequently overwinter on the ground where conditions may be more favourable for potential control agents than when the leaf is on the tree. Similarly soil-borne organisms surviving on debris are subject to the possibilities of various forms of competition with either ruderal, stress tolerant or combative organisms as described above.

So, as with all control measures (chemical, biological, integrated, cultural control or plant resistance) it is necessary to know the pathogen and the host before weak links in the life cycle can be identified as sites for possible control.

1.5   Host defence mechanisms

It is well known that diseases are often specific and one sort of plant may get a disease but another will not. Potatoes get late blight (caused by *Phytophthora infestans*) but oak trees never will: the oak trees are immune to this fungus. There are, however, more complex relationships than this all or nothing response, in which a fungus or bacterium is a pathogen on a species, but particular varieties, cultivars or individual plants vary in the amount of disease that they suffer. This is resistance in the host which is controlled by the host genome and it is the basis of plant breeding programmes to produce new crop varieties that are resistant to certain diseases (Nelson, 1973). Thus the black stem rust of wheat (*Puccinia graminis*) causes a serious disease of many species of grasses and cultivated cereals, though some species are more severely damaged than others. Even within one species, wheat (*Triticum aestivum*), there are varieties and artificially produced cultivars that do not get black stem rust; they are resistant to that disease.

Not only do hosts vary in their resistance but the pathogen varies in its virulence, so that a race of a pathogen may be able to attack this or that cultivar that does not have resistance to that race; the fungus is therefore virulent on those cultivars. Those same cultivars may, however, be resistant to different races of the pathogen. The situation, therefore, becomes very complex with cultivars or species resistant to some but not all pathogen races and pathogens virulent on some but not all cultivars.

The host plant has many methods of defence against the invasion by potential pathogens and these have been extensively described (Horsfall

& Cowling, vol. 5, 1980; Bailey, 1986; a simplified description is in most plant pathology texts or in Campbell, 1985; Fig. 1.12). There may be structural changes in the plant cell wall to lay down lignin or to form additional cross links between the structural carbohydrates which make it more difficult to degrade. There are changes in the host's metabolism which are manifested as an increase in ·the permeability of the

Fig. 1.12. The process of disease development looked at from the point of view of the potential host. (Modified from Campbell, R. (1985). *Plant Microbiology*. London: Edward Arnold.)

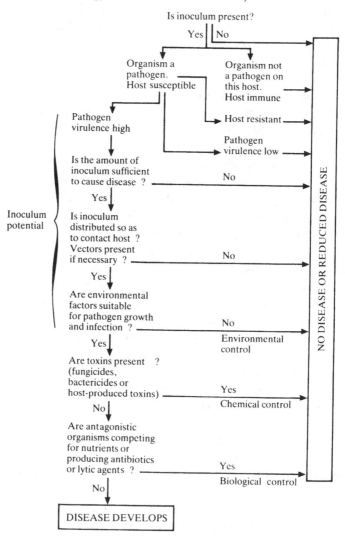

plasmalemma, a rise in the respiration rate, changes in enzyme levels and the accumulation of many new or altered compounds (phytoalexins) which may be toxic to the invading organism (Bailey & Mansfield, 1982). The plant may also have some of its cells killed very quickly in the hypersensitive response so that the potential pathogen, especially if it is a biotroph, is isolated from the live cells and is killed.

Not all plants have all these possible defence mechanisms and it should be stressed that the normal condition is a healthy plant, not a diseased one. Some of these defence mechanisms may be formed regardless of any infection but others are there as a response to a challenge by a potential pathogen. The challenge may be in chemicals produced by the invading organism, so-called elicitors, or in the detail of the chemistry or structure of the outer walls. This implies that there is a recognition system between the host and the potential pathogen so that it is identified as a danger and the defence mechanisms are triggered. Alternatively it is not recognized, or is considered a harmless sapro-troph, and so invokes no response. The recognized pathogen is the one that is not virulent on that host, the host defence works and the host is resistant. The virulent pathogen is the one that avoids being recognized as such, and so invades without provoking defences: the host is not resistant and disease develops.

The exploitation of these defence mechanisms in plant breeding programmes has taken place for many years even though the details of the biochemistry were, and to some extent still are, not fully understood. Some consider the plant breeding itself to be a form of biological control in that it is control of disease mediated by the plant, an organism other than man (Cook & Baker, 1983). Plant breeding has produced, by classical genetic techniques, cultivars with enhanced resistance to particular races of the pathogen.

It is also possible to use the information on the recognition system to confuse or mislead the defence. This is the basis of a form of biological control – induced resistance – in which the host defence system is challenged by a harmless organism, the biocontrol agent, but is made to react as though for a real pathogen so that the defence is already operating for any subsequent attack. This is the same sort of idea as vaccinating animals or humans against disease, but it must be stressed that the chemical mechanisms involved are completely different. The biocontrol agent must, therefore, be sufficiently closely related to the real pathogen, or otherwise have such similar chemistry, that it is mistaken for a potential pathogen.

There are several examples of such systems. Firstly, there is the use of avirulent forms or races of a fungal pathogen which, being so closely related to the real pathogen, occupy nearly the same niche. This may, therefore, lead to competition or it may be linked with the triggering of the host defence mechanisms. Thus inoculation, prior to planting, with avirulent *Fusarium oxysporum* can reduce the amount of wilt at later stages of growth. It need not be the same species of fungus, for in the classic work on this by Kuć (see Kuć, 1981; Horsfall & Cowling, vol. 5, 1980) the resistance of beans to anthracnose (*Colletotrichum lindemuthianum*) could be induced by heat attentuated pathogenic races or by the fungus *C. lagenarium* which normally causes the disease on cucurbits but not on beans. If the first or the second leaf is infected with the control agent, the next leaves which are formed, possibly over the following four or five weeks, are then resistant to the disease (section 3.6.2). It is not clear how this cross protection occurs: it is assumed that some chemical produced in the infected part is translocated to the new regions, and there 'conditions' the cells so that they respond more quickly or produce more phytoalexin, or whatever, when the real pathogen arrives, but there is little real evidence for this. Infection of the lower leaves of cucurbits with virus, such as Tobacco necrosis or with the bacterium *Pseudomonas*, also gave the new leaves protection against the fungal pathogen, even when the leaves to which the control agent was applied had now been removed from the plant. The signal for protection is host specific and graft transferrable but may be induced by many different factors. A further possible mechanism for induced resistance, at least in the wilt pathogens, is that the initial inoculation may induce tyloses that hinder the spread of the pathogen which arrives later: this is the case with *Cephalosporium* giving protection from *Fusarium oxysporum* f. sp. *lycopersici*. Whatever the mechanisms, and there may be several for the same or different combinations of pathogen, host and protecting organism, the control does work.

Inoculation with harmless viruses can give protection against pathogenic strains of this or other viruses which cause diseases. Pathogenic viruses may cause local lesions or they may be systemic, and the control may be by decreasing the number or size of the lesions (Table 1.2), by reducing the symptom expression or the number of new virus particles produced. In this case the protection may again be by the production of a translocated chemical, but there is also evidence that it may operate on a cellular level as one cell may not be infected by more than one virus, or if they are then only one of the viruses is reproduced. Some of these induced resistance mechanisms with viruses last many years, as for example when healthy citrus seedlings are inoculated with an avirulent strain of citrus tristeza virus.

Induced resistance to bacterial diseases can also be caused by other bacteria and here there is evidence that there are ultrastructural changes in the protected cells and that the initial stages in producing the immunity to later disease involves the recognition of the lipopolysaccharides on the bacterium surface (see Horsfall & Cowling, 1980). Infiltration of leaves with heat killed *Pseudomonas solanacearum* can give protection against the virulent pathogen and here some type of inhibitor is thought to be induced in response to the recognition of even the dead cells.

Linked with these examples of induced resistance are other less clear-cut examples of the use of the recognition and host defence mechanisms in biological control. Siderophores were mentioned above (section 1.3.2) and their prime activity is in competition via chelation of iron. There is, however, now evidence that they can also affect the production of phytoalexins, so changing specificity and resistance (section 6.3). The siderophore may well be produced by a saprotroph but it may trigger a response that affects a pathogen which arrives later.

Table 1.2 *The effect of systemic infection with a variety of viruses on the subsequent necrotic infection with cabbage black ringspot or potato X virus in tobacco*

	% reduction in number of lesions	
	Cabbage black ringspot	Potato X virus strain Br
Systemic virus, pre-inoculated		
Virus X: strain AST1	89	88
X4	78	28
AST4	76	0
Tobacco mosaic, type	96	85
masked	81	—
Potato virus Y	93	increase of 170
Tobacco severe etch	94	55
Cucumber mosaic	98	86

Based on Thomson, A. D. (1958). Reprinted by permission from *Nature* **181**, 1547–8. Copyright © 1958 Macmillan Magazines Ltd.

Finally in this consideration of the exploitation of defence mechanisms of the host there is the well-known case of *Agrobacterium* which causes crown gall on many species of plant and is virulent because of the possession of a plasmid which is transmitted to the eukaryotic host and expressed there via the host's metabolism (section 4.4). There is a specific recognition system based on the surface lipopolysaccharides and control can be obtained by using avirulent races or heat and ultra-violet killed cells that occupy the recognition sites and prevent the attachment and subsequent infection of the virulent form. In a specially selected avirulent strain of the bacterium the plasmid also carries DNA to code for a protein antibiotic active against other races of *Agrobacterium* (a bacteriocin, called agrocin 84). Commercially available biocontrol systems depend on using avirulent *Agrobacterium* with the ability to produce agrocin 84 (Table 1.3). Thus the specific binding sites are occupied and the antibiotic prevents most of the virulent strains from growing.

So this exploitation of the known host defence mechanisms is an area of great interest at the moment. As more becomes known it will undoubtedly be possible to find potential control agents with the correct wall chemistry to mislead the recognition systems, or with the correct chemical elicitors to initiate a metabolic response in the presence of a harmless saprophyte rather than the pathogen itself. Indeed it may be

Table 1.3 *Effectiveness of* Agrobacterium radiobacter *strain 84 in preventing crown gall in tomatoes inoculated with a variety of strains of* A. tumefaciens

Pathogenic bacterial strains	% galling with ratio of pathogen to strain 84 of:		Sensitivity of pathogen to bacteriocin 84
	1:1	1:10	
Q51	0	0	+
K27	0	0	+
K29	0	0	+
B234	0	0	+
EU 8	100	89	−
B6	100	100	−
Combined pathogen strains	89	83	

From Moore, L. W. (1977). *Phytopathology* **67**, 139–44.

possible now to use genetic engineering to put one or more of the presently known elicitor systems into existing organisms. Though such genetically engineered biocontrol agents would not be cleared for environmental release under the existing regulations, they may be allowed in the future when more is understood about the movement, stability and distribution of introduced micro-organisms.

1.6   Agricultural cropping systems and biocontrol

Many cultural practices affect disease (see Palti, 1981). Some result in mechanical disturbance and the mixing of pathogen inoculum with the soil or organic matter, some affect the water status or the physical properties of the soil, but some of the cultural practices may control disease by influencing the biological balance and can, therefore, be considered as biological control.

The most extensively recognized, traditional cropping system is the use of rotation of different crops. Rotations may be used for reasons of soil fertility or structure but if similar crops, which may have the same pathogens, do not follow one another then there is a good chance that any inoculum left in the soil will have died from starvation in the absence of its host or it will have been parasitized and lysed by other micro-organisms. This is not true of some diseases with very long lived spores such as *Plasmodiophora brassicae* which may survive up to 20 years in the absence of its brassica hosts. However, for many diseases the removal of the host for even one year (e.g. eyespot of cereals caused by *Pseudocercosporella*) will limit disease. Apart from using different species of host it may be possible to obtain some advantage by using rotations of different cultivars of the same crop, provided that the cultivars have resistance to different races of the pathogen.

Some of the advantages of the rotation system can be obtained with mixed cropping (Francis, 1986), growing two different plants together at the same time.

For some crops and some diseases there may be other strategies worth pursuing. Take-all, a root disease of cereals, can usually be controlled by rotations but this limits the crop choice for the farmer. If cereals are grown in monoculture (the same crop year after year on the same ground) then take-all increases for the first years as expected, but then the disease declines in importance, usually to an acceptable level. The disease stays at this low level as long as cereals are grown. A short break in the monoculture, for 1 or 2 years may have little effect on the decline of disease, but longer periods or certain types of break crops may destroy the decline phenomenon and the disease becomes worse again when wheat is replanted. This sort of decline is known from several diseases (e.g. scab of potato (section 5.3) and *Rhizoctonia* on radish) and the most popular theories, amongst many others, for these decline situations invoke some sort of increase in antagonistic soil flora in the continued presence of the pathogen (section 5.2.2).

This decline situation with particular diseases is a part of a wider phenomenon of suppressive soils which discourage disease. Suppressiveness may be associated with many soil characteristics (Schneider, 1982) such as particular clay fractions, but it is also destroyed by heating and is transferrable by a small inoculum to other soils. These latter properties suggest that the suppressiveness may have, at least in part, a microbiological origin (section 5.2).

Apart from the change from rotation to monoculture there is frequently a change in tillage system. Traditionally the land was dug or ploughed, then harrowed and/or rolled before seeding. This takes a lot of time and, because it involves several passes across the field, the machines use a lot of fuel. Moves to economize on fuel have lead to reduced tillage or minimum tillage or even no tillage (direct drilling) in which seed is sown directly into the remains of the previous crop. This has several results: it is often cheaper and certainly quicker, so it allows earlier sowing but in heavy soils it may cause compaction. There is a surface build-up of crop residues that are not ploughed into the soil and this can be deleterious because their decay causes oxygen and nutrient shortage to the germinating seeds and the developing crop. However, the surface cover may be useful in reducing water loss in arid-land agriculture. No tillage may be combined with burning of the crop remains to reduce the surface build-up of organic matter. This is obviously a complicated subject with many ramifications (Cook & Baker, 1983; Lynch, 1983; Campbell, 1985) and though the effects on disease, which depend on the soil and the climate, may be understandable they may not be predictable. In dry conditions the pathogens from the last crop may survive on the residue and disease control is improved if this is ploughed into the moist soil at depth so that it may be attacked by antagonists. In wetter conditions decay and pathogen control may take place on the surface of the soil and ploughing-in a residue may simply place the diseased remains in contact with the roots of the developing crop. There may, however, be confusion of this interpretation by the crop residue acting as an organic amendment (see below) and by changes in drainage, soil aeration and other physical factors that the ploughing or other soil disturbance causes. Again many factors are operating simultaneously and it may be difficult to determine which are most important in a particular case.

The effects of organic amendments and green manures (section 5.3) added to the soil have been claimed to be beneficial and are one of the main ways by which nutrients are added in organic farming. Traditionally this involves the addition of farmyard manure after a period of composting to reduce the carbon to nitrogen ratio. Green manuring is the ploughing-in of a crop like alfalfa (also called lucerne, *Medicago sativa*), which may have been grown specially for this purpose. There are clearly several possible modes of action for the addition of organic matter: it can supply nutrients to the general soil population and the plant, it can add a specific substrate, or it can be used to improve soil structure, though very large quantities will be needed to produce any long-term change in the humic and fulvic acid fractions which compose the major part of soil organic matter. The rise in readily decomposed soil organic matter on addition of the material is transitory, and most is decomposed quite quickly, by increased microbial activity, in a matter of a few weeks or a year or so. Improved crumb structure can increase drainage and aeration in heavy soils and conversely may improve water retention in sandy soils. In the process of decomposition there may be immobilization of nitrogen and a depletion of the oxygen in the upper horizons of the soil, especially if the material is just left on the surface in minimum tillage systems rather than being mixed in by some form of cultivation (Lynch, 1983; Campbell, 1985). Low oxygen and high carbon dioxide can themselves inhibit pathogens such as *Rhizoctonia*.

The addition of organic matter can, therefore, supply specific substrates for particular organisms and affect the aeration and the drainage to improve plant vigour. However, the most important effect of organic additions is to increase the general level of microbial activity by breaking the dormancy imposed by carbon and nitrogen limitation. The more microbes that are active the greater the chances that some of them will be antagonistic to pathogens. This general response to organic matter with a reduction in pathogen inoculum (Fig. 1.13) has been used for many years to control such diseases as potato scab (*Streptomyces scabies*), *Phymatotrichum omnivorum* rot, and the sclerotia of *Sclerotium rolfsii* and *Rhizoctonia*.

The level of available nitrogen in the residue may have a significant effect on the response of the pathogen and the microflora. Low available nitrogen may favour the pathogen by allowing it to maintain possession of a substrate in the absence of much activity from competing organisms (see above). This is the case with organisms using straw as a food base (e.g. *Cephalosporium, Pseudocercosporella*) or growing on timber (*Armillaria mellea*). Alternatively higher nitrogen levels may favour survival of the pathogen (Fig. 1.13) by allowing its continued saprophytic activity where lack of nitrogen can lead to starvation. The control of nitrogen levels may be determined by the substrate itself, by the decisions on the application of fertilizers or by manipulating the cropping system. For example, undersowing wheat or barley with legume/grass mixtures has been used to control take-all (*Gaeuman-*

Fig. 1.13. The inoculum density of *Thielaviopsis basicola* as affected by amendments added at various intervals before assay. Inoculum decreases with time anyway (control), glucose has little effect, ammonium nitrate increases inoculum survival and alfalfa decreases survival especially in the early, microbiologically active stage of decay. The two soils responded slightly differently, especially to nitrogen. (From Papavizas, G. C. (1968). *Phytopathology* **58**, 421–8.)

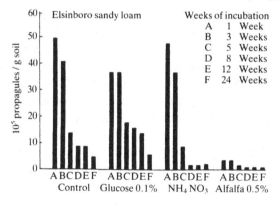

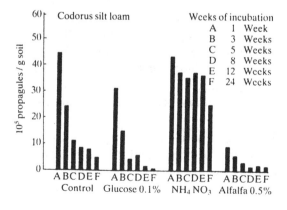

nomyces graminis). The undersown crop may supply some nitrogen by biological fixation but after harvesting the cereal and during the autumn the nitrogen is immobilized, and therefore conserved, in the growing cover crop and nitrogen starvation decreases the activity of *Gaeumannomyces* (Garrett, 1970). The legume mixture may then be used as a green manure before sowing barley, and the decomposition immobilizes nitrogen until spring when with the rising soil temperature the decomposition is completed and the nitrogen released. An undersown crop may also give soil cover, which may be important in erosion control. There may be other factors operating here, for grass leys encourage a fungus called *Phialophora* which is an antagonist to take-all, and legumes can encourage the growth and persistence of bacterial antagonists. Biological control mechanisms, especially in soil, are rarely simple.

The form of the nitrogen, as well as its availability, may influence the microbes and the pathogen. For example, the germination of sclerotia of *Sclerotium rolfsii* is inhibited by pH changes caused by ammonia and by microbial antagonism encouraged by nitrogen supplied as urea, ammonium, calcium nitrate, peptone, etc.

Minimum tillage, which has been especially used in monoculture of cereals, allows earlier sowing of winter crops because less time is spent in ground preparation. This gives increased yield but because the plants are put back on the ground so soon after harvest there is an increased risk of disease carrying over from the last crop. The vigorous growth in the warm autumn soil also leads to a nitrogen requirement and the lush growth is susceptible to leaf diseases. More fungicides may be needed than when seed is sown in late autumn or early winter. All the factors, therefore, interact to affect the levels of disease including rotations, monoculture, the length of break between crops, the tillage system, nitrogen level and type, crop growth stage, plant resistance and pathogen survival.

This has been a very brief overview of the ways in which biological control may operate and of how the farmer may manipulate plant resistance, cultural practices and tillage systems to favour particular organisms or particular conditions. It has by no means been exhaustive, but is intended to serve as both an introduction and a background so that the reader will recognize what may be occurring, and realize at least some of the more general implications, when more detailed examples are discussed in the following chapters. It is important to stress that these different organisms and control systems will operate together in an agricultural situation (even though they are described or studied separately). Biological control requires the management of the whole agro-ecosystem, in terms of crop rotation, fertilizer or organic amendment addition and introduced antagonists: because biocontrol involves dynamic, yet quite stable, equilibria between different organisms and the environment, the overall situation must be studied to shift the equilibria in the desired direction.

2

Historical and commercial background and methodology of biological control

2.1   Historical background

Biological control, in its widest sense, has been used by man almost since the beginnings of organized arable agriculture. In the third and fourth millennia BC, fallowing, limited forms of crop rotation and mixed or inter-cropping were used in the fertile crescent of the Middle East and later by the Chinese. More recently the Saxon and medieval field systems of Europe also used simple rotation and fallow periods to try to reduce disease and increase fertility. The term biological control was coined in connection with the control of insect pests by the introduction of predators (Howard, 1916, Smith, 1919, in Cook & Baker, 1983; Baker, 1987). The use of non-pathogenic micro-organisms to control plant disease occurred at almost the same time, but was not specifically called biological control. Mechanisms of action were also being investigated, and in 1928 Fleming discovered a particular antibiotic, penicillin, which was later isolated, purified and chemically character-ized by Florey, Chain and co-workers.

In the 1920s there was a sudden increase in the number of publications reporting the control of disease by antagonistic fungi, actinomycetes or general soil populations. Thus in 1921 Hartley introduced antagonistic fungi, isolated from soil, to control damping off in pine seedlings (Fig. 2.1) in partially sterilized soil, though the effect was apparent in field soil. There are complications in the modern interpretation of this experiment because there were obviously toxicity problems due to sterilization, and the added saprophytic inoculum did give growth improvement in the absence of the pathogen. Sanford (1926) and Millard & Taylor (1927) both showed control of potato scab (caused by *Streptomyces scabies*) to be connected with the activity of antagonistic microbes, which were encouraged by green manures.

Sanford used isolates of bacteria from soils to investigate the interactions in culture and considered that the inhibition of scab was mostly due to acidity produced by bacteria such as *Bacillus (=Pseudomonas) fluorescens*. There were, however, some bacteria which inhibited the *Streptomyces scabies* but which did not seem to produce acid and Sanford noted that 'the phenomenon is not unlike the activity of a lytic principle . . . When scab is controlled . . . it is suggested that the antibiotic qualities of certain predominant soil micro-organisms influence the development of *Actinomyces [=Streptomyces] scabies*'. Millard and Taylor isolated *Actinomyces praecox* from soil treated with green manures and showed that scab could be controlled by inoculating *A. praecox* back into infested soil.

Fig. 2.1. Pine seedling survival in the presence and absence of *Pythium*, with (⸺ B & D) and without (--- A & C) antagonistic saprotrophs (see text). The saprotrophs increase survival. (From Hartley, C. (1921). *Bulletin 934, USDA*. 49 pp.)

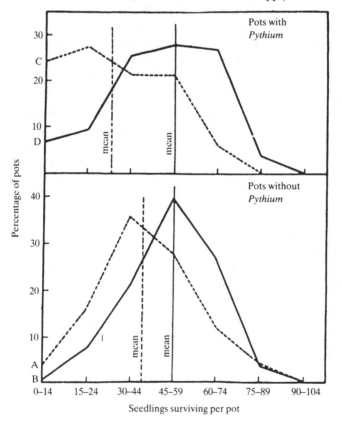

In 1929 McKinney demonstrated the induced resistance effect with viruses, which was discussed in the last chapter (section 1.5), though the observation was passed over as incidental, without its significance being recognized.

The control of fungal root diseases of cereals by resident micro-organisms and by the general soil population was shown by Henry (1931) in a series of simple experiments (Fig. 2.2).

In just ten years there was, therefore, a demonstration of the control of fungal, bacterial and viral diseases of seedlings and mature crops, with many of the main ideas and suggested mechanisms of action that are still the basis of biocontrol research 60 years later. There was then a continuous, though slow, supply of reports expanding the information to different crops and diseases and also suggesting more modes of action of the introduced or encouraged micro-organisms. These reports were interesting scientific curiosities that really evoked no serious discussion, and until the 1960s they did not lead to any commercial use of biological control agents against plant diseases. Then, in 1963 Rishbeth published his paper on the control of *Fomes* (now *Heterobasidion*) *annosus* by *Peniophora* (now *Phlebia*) *gigantea* and this became a commercially available product, though marketed in quite a small way to a specialized market. The *Peniophora* colonizes tree stumps and competitively excludes the *Fomes* (see section 4.2.2). There followed the first international conference on the use of biological control (Baker &

Fig. 2.2. Relative degree of foot rot infection of wheat seedlings caused by *Helminthosporium sativum* with various soil saprophytes. (From Henry, A. W. (1931). *Canadian Journal of Research* **4**, 69–77.)

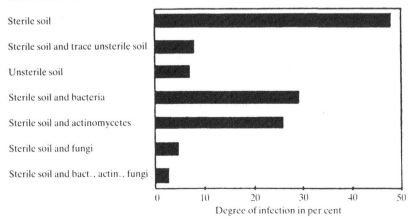

SUBSTRATA OF PATHOGEN

Snyder, 1965) which was the forerunner of a series (Toussoun *et al.*, 1970; Bruehl, 1975; Schippers & Gams, 1979; Parker *et al.*, 1985). Also during this time the two most important books on biological control were published (Baker & Cook, 1974; Cook & Baker, 1983) and there were a number of reviews and shorter texts (Corke & Rishbeth, 1981; Deacon, 1983; Papavizas, 1985). More commercial products were also marketed at this time, especially those based on the fungus *Trichoderma*, which is presently the most widely used control agent with different species and strains available for the control of a number of diseases.

There are therefore three main phases in the historical background to biological control. Firstly, the prehistoric and traditional use of crop manipulation which may lead to control by biological means. Secondly, the steady growth of information for some 60 years of this century, which has in the last 20 years greatly increased in volume. Thirdly, there is the present situation (1988) of extensive research in many parts of the world (with many different diseases being studied) which is for the first time receiving reasonable levels of funding from the government research agencies and from commercial interests. Why is there so much interest in biological control when only a short while ago it was but a scientific curiosity, and why has it taken so long to realize the significance of the original work?

2.2   Recent interest in biological control

One of the fundamental reasons for this interest is that we now have a sufficient, though by no means complete, knowledge of microbial ecology and plant pathology so that there is some chance of understanding the niches which micro-organisms must colonize and the microbial interactions involved. This also allows some degree of prediction, so that suitable diseases for biological control may be selected for study and likely courses of action can be considered: the research has some direction and logic. New developments in genetic engineering also allow microbes to be changed and adapted to make them better inocula or more efficient in fermentation systems. It should also be possible to combine the most desirable characteristics of several organisms in one agent that will have several mechanisms of attacking a pathogen, as well as good survival and colonization characteristics (section 5.1). So the means are now at hand, but we still need to know what are the main driving forces to develop the biological control of plant pathogens.

The success of chemical pesticides in controlling plant diseases (and insect pests) has worked against biological control. While there were very effective, relatively cheap methods available there was no incentive for the development or marketing of other systems and the problems which are considered important today, such as the concern about the environmental effects and safety of chemical pesticides were not fully recognized. Certainly some of the early pesticides were potent toxins (e.g. mercury and organochlorine insecticides). They persisted in the environment, accumulated in predators at the top of food chains and were shown to have long-term effects on non-target organisms. The use of such chemicals has now been discouraged or prohibited in many countries, though they are in use in some places where they still give cheap, effective control of some pests and diseases, despite the environmental damage they cause. Modern pesticides have to pass very stringent tests for safety and for lack of any environmental hazard. The fact remains that they are toxins and occasional examples of misuse or unexpected side effects do occur. It is estimated that there are about 3000 hospitalizations and 200 fatalities per year due to pesticides, apart from problems which are unrecognized or not considered serious enough to warrant medical attention (Pimental *et al.*, 1983). About half a million tonnes of pesticide are used annually and yet one third of all crop production is still lost. If the use of pesticides was prohibited there would be even worse disease problems in animals (including man) and in crops. The latter would result in food shortages on a world scale, for all the highly efficient and very productive western agriculture is based on the use of fertilizers and pesticides of one sort or another. So the immediate banning of pesticides is not possible, even if it were considered necessary. There is however a longer-term move by 'environmentalists', some political parties or pressure groups and many responsible agronomists, plant pathologists and ecologists for at least a reduction in the use of pesticides and more effective codes of practice or legislation to control their use. There is a need to find other ways of controlling plant diseases and these will include all the methods of crop rotation, plant breeding, etc. which have been discussed and also the use of biological control alongside limited chemical methods in integrated control programmes (see sections 2.9 and 5.9). This gives the possibility of a commercial market in biological control agents for plant diseases and has resulted in the interest which agrochemical companies are now showing in the development of such products.

There are also other commercial reasons for the recent interest. If we take the case of fungicides as an example of a pesticide, they are discovered by an essentially random process of testing as many chemicals as possible for effects on fungi. A large agrochemical company may test several tens of thousands of chemicals each year and the hope was that one in about 5000 might lead to a product. This one active ingredient would then be formulated in several different ways (different strengths, powders or liquids, in combination with other fungicides, etc.). Once having found an active chemical group it is also worth looking more systematically at related compounds or chemical modifications of the original substance. There is still a lot to be done, even with the existing chemicals, to make them more effective at lower doses, more easily degradable in the environment and so on. However it has recently become clear that it is now more difficult to find new compounds; perhaps only 1 in 15 000 tested now becomes a product. This is partly because new compounds have to pass more stringent tests than previously, but there is also the suspicion that perhaps most of the effective chemicals have already been found. Fungi develop resistance to fungicides so new chemicals are constantly needed, but they are becoming more difficult, and therefore more expensive, to find (Delp, 1977; Lewis, 1977; Campbell, 1986). Costing biological agents has not been done very extensively, or at least such information has not been published even though it has formed part of the assessment made by commercial interests in biocontrol. However it is likely that development and registration costs will be less than chemicals and some of the existing biological control agents, especially against insect pests, are very attractive economically in terms of value of crops saved in relation to cost. Whether there is commercial profit in their sale may be another matter. Care is needed because initial studies show that the factors governing the long term returns on biological control are different from those for chemical pesticides.

It should not, however, be assumed that biocontrol agents are the answer to all the problems. There are many reasons why they have not been used in the past and why there will be problems in the future. Firstly, there are still other cheap and very effective control measures for many diseases: leaf diseases are controlled by breeding resistant plants and by a very wide range of good fungicides. The question, which it will take a long time to be answered, is whether biological control could have reached the same degree of success as chemical control has, given the same amount of research time and finance. Biological control is making a good start, but it is decades behind chemicals in its development programme.

Another problem with biological control is the difficulty of introducing a 'foreign' organism into a complex environment such as soil.

Those control agents that are successful now tend to operate in environments without competition such as virtually sterile horticultural composts, fumigated soil or in clean timber (Corke & Rishbeth, 1981; Papavizas, 1985). In more complex environments the colonization, and therefore the control, is patchy (Baker, 1986).

It must also be remembered that biological control agents themselves are not exempt from restrictions that are imposed on chemical pesticides. Biocontrol agents must be safe: this not only means that they should not be human or animal pathogens, it also implies that they will not deleteriously affect the plants or natural soil populations, or get into water supplies or spread to other environments where they could be a problem. These basic requirements are probably more complicated to test for a biocontrol agent than for a chemical, for the control agent could grow, reproduce or genetically change itself or other microbes after it has been released. There is no *a priori* reason why biological control should be safer than the use of chemicals and new considerations apply to risk assessment criteria developed for pesticides (Alexander, 1986). There is no real reason why micro-organisms selected for agricultural use, or even genetically engineered, should be any more unstable or dangerous than the laboratory strains that have been randomly mutated, selected and used by industry, without special containment, for many years. *Rhizobium, Azotobacter* and *Bacillus* are already widely released in agriculture without any serious problem having developed (Brill, 1985). Care is clearly needed, but there seems little evidence that the extreme fears expressed by some are justified.

Not only must biocontrol agents be as cheap and as safe as chemicals, they must also be as easy to use. There are again problems here associated with being a living organism that may have a limited shelf life, and be sensitive to changes in temperature or osmotic pressure. There is a very great difference between the use of a biocontrol agent in the sterile laboratory with skilled microbiologists and its use on a farm with no special facilities in poor weather conditions. You can no more expect the farmer to use complex microbiological procedures in storage or use than you can assume that he or she will synthesize some organic chemical to use as a pesticide! Farmers will only use difficult procedures with expensive and unreliable control agents if there is nothing else that is legally available, so it is the job of the expert to make a biocontrol agent which the farmers want to use because it can effectively, cheaply and easily control some disease which is of importance to their crops.

2.3   Isolating biological control agents

The aim of the investigation must be decided at the start (Fig. 2.3). If there is to be an enquiry into some particular sort of antagonism then a scientifically interesting set of organisms should be obtained which show this characteristic to the best advantage, and it does not matter if they grow well or survive under normal farming conditions; whatever it is that they do it will occur under controlled laboratory conditions. If the aim is to produce a marketable product then a whole different range of commercial considerations will be the first concern (Scher & Castagno,

Fig. 2.3. A generalized scheme for the isolation and study of biocontrol agents.

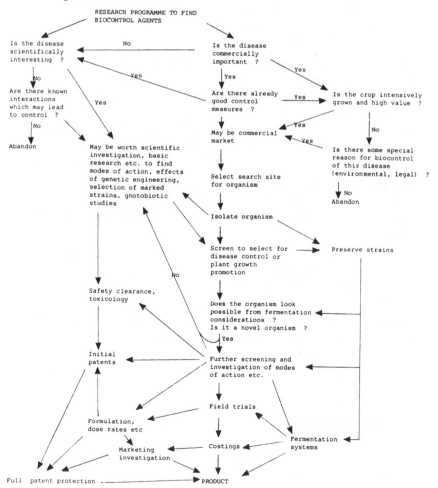

1986) and many scientifically interesting organisms may be discarded in the search for a suitable, commercially viable control agent.

The next stage is to decide what sort(s) of diseases are most suitable for this aim, and then to work out a rational way to search for appropriate organisms. Firstly, for commercial use the disease must be important, either for economic or aesthetic reasons, so that if a control agent is produced people will use it. More correctly, it is commercially important that people will buy the product, then use it. Research for the biocontrol of diseases of major crops (cereals, potatoes, vines, soybeans) in intensive western agriculture is in progress because the farmers have the money to buy such a product. Crops important in subsistence farming will not provide a commercial market, though there may be many methods of biological control that are suitable and available. It is notable that rice (certainly a major world crop) has had comparatively little attention (see section 5.6.1), though pests like the brown rice hopper are under study. Rice is predominantly grown in the developing countries rather than in Europe and North America.

Secondly, if the host is a high value crop, grown intensively, then there is more latitude in pricing, as such a crop can absorb extra costs with comparatively little effect on the profit. Also it is easier, theoretically, to control diseases of crops which grow and mature quickly so that the survival time of the control agent in the environment is at a minimum. These factors point to horticultural crops and glasshouse crops as likely targets and this has the further advantage that there is the possibility of controlling the environment to favour the antagonist and of using modified, possibly even sterilized, composts so that it is easier to introduce antagonists into normally complex environments such as soil. A similar case can be made for the control of diseases of cultivated mushrooms which are a high value crop, grown very intensively on specially prepared composts. They suffer from a variety of diseases and pests, including bacterial brown blotch (caused by *Pseudomonas tolaasii*) which can be controlled by applying antagonists, such as other species of *Pseudomonas*, to the casing soil to prevent the build-up of disease.

Thirdly, since the proposed control agent will have to compete with other methods, the disease chosen should ideally not have any simple, effective system of control already in use. If that is not true then there should be some obvious advantage that a biocontrol agent may be expected to possess. This might be greater convenience, increased safety or greater environmental acceptability. There are so few biocontrol agents already available that there are many diseases which fulfil these criteria, and many where a search for biocontrol agents would be less desirable at present. For example there are so many ways of controlling many leaf diseases (chemicals, plant breeding) that only some very special biocontrol agent would offer any advantage: why try to search for the special when we do not even have many ordinary biocontrol agents! The corollary to this is that root diseases, because they are difficult to detect, assay and treat chemically, have comparatively few existing control measures and may be suitable subjects for biocontrol agents.

These considerations apply to the place of control agents in the complex agriculture of the so-called developed world. In Third World countries it may not be that other control agents do not exist, but rather that they cannot be bought by the poorer farmers. What might be needed here is a control agent that can be produced and used in a low technology society. Such systems as the use of composts or green manures to enhance or produce biocontrol agents may be applicable, even though they are labour intensive. In high technology agriculture they would be less suitable because the labour and transport costs in such societies are at present too high to allow their economic use in competition with commercially produced agrochemicals.

So we have a suitable disease which for a variety of reasons may be considered as a target for biocontrol. We now need to know as much as possible about the disease and its survival strategies etc., as discussed in the last chapter, so that we can decide on the sort of control agent to be used. This should normally be associated with the host plant of the disease in question, or in the soil or district in which it grows, so that the control agent eventually selected will be able to survive and grow in the environment in which it is expected to operate. The search sites for organisms should be subject to the normal agricultural practice that will be experienced, so that the organism is again tolerant to the pesticides, fertilizers, etc. that are in use.

The particular disease conditions at the sampling site are important. Serious outbreaks of disease are not usually good places to look for control organisms, for if they were there then the disease would not be serious. More suitable are conditions where the disease would be expected to occur but where it is absent, or at least not serious. This lack of disease should not be explained solely by environmental factors, but should be caused by rotation, crop mixtures, soil conditions or whatever that are known to help in the control of the disease. Any plants or patches of plants that are healthy in an otherwise diseased site would be worth investigating.

The places to look for potential control agents must be selected carefully. Random contaminants of laboratory cultures, or even isolates from culture collections, rarely produce useful organisms for the field as the micro-organisms in laboratories are usually adapted to the high nutrient levels in common media. A great deal of time and money are going to be invested in the development of the organisms so a little consideration at the start is well worth while.

2.4   Development of commercial strains

Having decided on the disease and the search site (Fig. 2.3), there are several possible strategies for the isolation of potential biocontrol agents and the methods have been described in detail (Andrews, 1985; Dhingra & Sinclair, 1985; Campbell, 1986). The methods must be selected from the start to meet the aims of the investigation. Genera or higher taxa known to be useful antagonists may be specially chosen by the use of selective media (e.g. for *Pseudomonas* and many fungi and actinomycetes) or by such procedures as pasteurization for *Bacillus*. If there are no known organisms that might be useful then more general media, usually very dilute, may be used: it is important that isolates are not adapted to the high nutrient conditions of normal laboratory culture if they are eventually expected to survive and grow in the wild.

These media may be used in standard dilution plates, the colonies being picked off and screened for potential activity as soon as possible. Alternatively the pathogen is seeded onto the plate which is then spread with dilutions from soil, roots, leaves, etc. and only those colonies which show antagonism, as detected by the inhibition of growth of the pathogen already on the plate, are selected. Another possibility is to use agar containing chitin, cell wall suspensions or other likely targets of lytic enzymes and to isolate organisms producing clearing zones. These methods which combine isolation with some degree of preliminary screening have the major disadvantage that the investigator prejudges the mode of action which is important, and so misses organisms that show growth promotion of the plant for example, but which do not give inhibition or clearing zones.

It may be considered worth while to select fast growing organisms or those that grow best on cheap media so that at later stages in the development there will be no problem with finding economic fermentation systems. These characteristics may however be thought a fairly minor problem that can be dealt with later, by standard methods of strain improvement or even by genetic engineering, the main aim of the initial isolation being to produce organisms that can grow in the much less understood and less controlled proposed environment. For example amoebae, which are known to parasitize pathogen hyphae (section 5.7), are not favoured control agents at the moment, mostly because there are no good systems for their mass growth (section 2.6). It will be much easier to justify commercial development if the investigation stays within, or close to, existing technology (Scher & Castagno, 1986): there are going to be enough problems at the field testing stage (section 2.7) without trying to develop brand new fermenter technology at the same time.

Looking even further ahead it may be thought desirable to isolate spore forming organisms so that they have a long shelf-life as a product and are easy to distribute to the farmer.

None of the above discussion applies to the use of biotrophs as biocontrol agents. Mycorrhizae are known to affect the development of root diseases (section 5.4), but there is no clear way in which vesicular arbuscular types may be isolated for they cannot be grown easily in culture. It is possible to develop theoretical systems involving growing the biotroph on a host and using this in conjunction with the pathogen as a trap for potential biocontrol agents. Such systems have not however been used: again there is so much still to be done with the simple, more manageable systems that there is no need to attempt the difficult or impossible yet.

After isolation the investigator has a collection of organisms, maybe containing hundreds of taxa and thousands of strains. It is most important that they are stored as soon as possible in a way that gives the minimum opportunity for mutation or selection of laboratory adapted strains. Normally the cultures are lyophilized, or stored in glycerol at $-70\,°C$ or in liquid nitrogen.

2.5   Screening potential control agents

There are two basically different systems: *in vivo* tests involving the whole diseased plant and *in vitro* tests using some sort of laboratory culture.

The *in vitro* tests are at first sight attractive: there is a clear, visible result (Fig. 1.9) such as the inhibition or lysis of the pathogen, they are relatively easy and quick to perform with large numbers of isolates and they give quantifiable data that are capable of numerical and statistical analysis. They are suitable for selecting organisms with a particular mode of action, but they are very poor predictors of the activity of the organisms in the field. The most widely used test is that to identify antibiotic producers (Dhingra and Sinclair, 1985; Andrews, 1985) in which the pathogen is inoculated onto an agar plate and the potential antagonist is point inoculated or streaked nearby. The degree of inhibition of pathogen growth, in relation to growth in the absence of the potential control agent(s), is used as a measure of effectiveness. The other *in vitro* method of assessment is to study interactions in slide cultures, but this is very time consuming. The use of *in vitro* methods may be unavoidable because of space limitations, the absence of the host plants and means of growing them in glasshouses (e.g. diseases of mature trees), or limited time and staff may prevent the use of an *in vivo* test.

In vivo tests are used for choice as they most closely imitate the conditions under which the control agent will eventually have to operate. A host plant, infected with the pathogen to be controlled, is used as the test organism, the presumed control agent is applied to the leaves or roots and after an appropriate incubation the amount of disease is compared with an unprotected control or with a healthy plant. The amount of disease control is measured regardless of the mode of action (Andrews, 1985; Campbell, 1986). There may have to be some simplification of the system with the use of seedling plants so that the test does not take too long or take up too much space. For root diseases sand or artificial composts may be used to get more uniform growth conditions, for it is very difficult to get large amounts of soil of a constant type for a big screening programme which may last several years. Similarly the plants may be grown in glasshouses or growth cabinets to control the day length, humidity etc. Special growing containers have also been used in an attempt to improve control over the degree of infection by the pathogen against which the control agent is to be tested (Weller, Zhang & Cook, 1985). The further removed the system is from the natural growing conditions the more reproducible the screen, but the poorer the predictive ability of final performance in the field. The overwhelming problem with *in vivo* screening is the time, effort, growing space and money that it takes. It may be possible to use a 2-stage screen with a small growing container under controlled conditions for the initial selection and then a larger pot or plant with more replication for a second test of the most promising organisms.

The screen involving the whole plant may test various other parameters that are involved in the ability to control disease, either as part of a single screening process or as a separate operation. Does the agent work by affecting inoculum in soil, by protecting the plant from infection or by curing the infection once it has occurred? If the pathogen and antagonists are mixed for some time before the host is introduced then you will test for reduction in survival of the inoculum. If the antagonist is pre-inoculated on the host then protective ability or competitive exclusion will be tested and if the infected plant is inoculated with the biocontrol agent the test will be for the cure of an existing infection. It may be possible to get complete cover of leaves by spraying and in some horticultural crops, which are transplanted, it may be possible to dip roots in the antagonist culture: colonization may not then be needed. If the antagonist is only placed discontinuously, or only on a seed for control of a root disease then it will be necessary to have growth and colonization (section 1.3.1) and this can be tested as a part of the screen. Those organisms which fail to colonize give no control and are rejected, even if they have the potential for control in the presence of the pathogen.

In any of the different sorts of screening it may be desirable to test more than one antagonist at a time, for some antagonists antagonize each other as well as the pathogen and others act synergistically. The final product containing more than one organism may have an extended range of soils or climatic conditions which can be tolerated. Again tests of multiple inoculants are relatively simple in culture, but much more difficult with a live plant test because the number of possible combinations of organisms requires many controls and the size of the experiment soon gets out of hand.

2.6   Testing selected antagonists

Even though the sequence of operations outlined above should have resulted in a collection of suitable organisms that show some sign of controlling the chosen disease, there is still a long way to go before we have a biological control agent. Of the original isolates there may be only about 1% left, indeed if there is more than 1% there will be problems in the next stages which are very labour intensive and often expensive: from now on it is just not possible to handle large numbers of different organisms. The problem is to find out how, when, where and under what conditions the selected antagonists work. There may already be indications of this from the screening, but detailed investigations are now needed (Cook & Baker, 1983; Baker & Cook, 1974; Dhingra & Sinclair, 1985; Campbell, 1986; Scher & Castagno, 1986) to provide information and materials for later field trials that will lead to the selection of an effective biocontrol agent.

The organism(s) must be identified so that it can be referred to in patent applications and literature searches can be made to discover what is already known of its activity. Names are usually also required for toxicology and safety clearance (Fig. 2.3). It is relatively easy to put a name on a fungus because this is done on morphological characteristics and fungi from natural environments are fairly well described in the literature. With bacteria there are problems, and often it is just not possible since so few bacteria from natural environments have been adequately characterized; most identification systems and most bacterial taxonomy is heavily biased towards medically important species or strains. Furthermore, the bacterial taxa are the subject of considerable controversy. There are, however, standard sets of biochemical tests to characterize a bacterial isolate and these generally lead to a generic designation and sometimes an approximation to a species. Almost invariably the diagnosis reads 'it is close to species x, but is atypical in that . . .'.

Safety clearance is at the moment a difficult area, for there are no obvious, agreed criteria which are based on sound information (Alexander, 1986). It goes almost without saying that human, animal and serious plant pathogens must be eliminated, and even opportunistic mammalian pathogens are very suspect because the organisms will be handled in large quantities in commercial use, which may pose problems. Most laboratories, both research and commercial, have their own house rules which are usually based on reasonable guesses about what precautions are necessary with skilled personnel. Once outside the laboratory microbiological skills cannot be assumed and different countries have different rules (or no rules at all!) about the release of organisms into the environment; it is just not known how long they might survive, where they might go or what they might do.

Any organisms that have been genetically engineered are, rightly in the present state of ignorance, subject to strict controls on release. Like all major innovations, such as pesticides and fertilizers, genetically engineered organisms may do enormous good, but they may also present some problems. Most genetically engineered organisms for biological control would be likely to have had several changes made (Lindemann, 1985): they would therefore be unlikely to have occurred naturally and we do not know how they would behave. Furthermore they have probably been selected or designed to survive in the environment, which is not good for an unknown introduction; indeed it is equivalent to creating some of the problems that occurred with the persistent pesticides like DDT. We have to ask, should a genetically engineered organism be released, if it is will it survive, if it survives will it grow and spread, if it does spread will it be detrimental and will the new genetic information be passed on to other organisms which may themselves grow and spread (Alexander, 1986)?

In general, organisms isolated from natural environments and simply returned there are permitted. But what about an organism from a natural environment that has been selected for several different characteristics? It may be just as different from the wild type as a genetically engineered organism, indeed in some ways it may be more dangerous because you usually do not know what exactly you have favoured or suppressed by your selection, whereas a genetically engineered microbe will have had a particular enzyme or whatever added or taken away. It seems likely that biocontrol agents will have to pass many of the environmental tests which pesticides undergo, though some are already in use without this. These matters are the subject of intense discussion on a national and international basis and some set of agreed rules will be drawn up: the problem is that the methodology for isolation, testing and tracking of particular strains of micro-organisms as they move in the environment is not easy, or even does not exist in a sufficiently developed form to give the answers that environmental protection agencies are demanding.

There is now a considerable technology associated with the design of delivery systems (Fig. 2.3) for micro-organisms to be used in agriculture, based originally on the long standing use of *Rhizobium*. The first question is where the organism is to be applied. Should it be direct sprayed on the leaf or applied by dipping the root, should the seed be coated with the antagonist(s), should they be applied to the soil and if so before, at, or after sowing? The mode of action of the antagonist may indicate that it actually needs to be in contact with the pathogen, or it may be able to give protection to the plant at long range; this will again affect the positioning of the antagonist and therefore the delivery system. There are several basically different methods (Papavizas, 1985). The agent may be supplied as a live liquid culture; while this is suitable for experimental work it is useless for any commercial inoculum. Dried, usually freeze dried, cultures may be supplied to be made up with water to a spray suspension in the field, though there is usually considerable loss of viability. The inoculum may be supplied as a powder or granule in some sort of carrier that is often finely ground peat, bran etc., sometimes with added nutrients to encourage growth of the antagonist, though hopefully not the pathogen. There are now a number of ways of encapsulating microbes in gels, for example those made of alginates, to produce powders or granules whose solubility, degradability and water retaining capacity can be carefully controlled. Whatever system is used the aim is to preserve viability of the micro-organism while allowing convenient handling and distribution to the correct place. A lot of testing is involved to discover the exact details of the best system: quite small differences in formulations and water content can greatly affect survival.

Linked with the development of delivery systems there will need to be tests on the effects of inoculum density of the antagonist, possibly at different inoculum potentials of the pathogen, to determine the amount needed for control. This will also include fermenter studies to discover the best way of mass producing the organism, while maintaining its effectiveness. It is unfortunately all to easy to grow micro-organisms in a fermenter to very high densities, but then to discover that you have selected a fermenter-adapted strain from the original inoculum, which grows very well at high nutrient levels, but does not actually work as a biological control agent. It is desirable to have micro-organisms that are stable genetically so that they maintain the desired properties through all the testing and production, but quality control and testing is invariably needed at all stages (Scher & Castagno, 1986). These production factors will have a very large bearing on the economics of the final control agent and if satisfactory answers are not found at this stage the organism may be abandoned, even though it seems to work quite well.

These then are the far distant aims of the detailed investigation of the selected control agents which must now be done, but you cannot select or improve strains to meet the above needs unless you know how they operate as control agents against the chosen pathogen, and therefore in what ways they should be better.

Ways of testing for the modes of action for the antibiotic producers and those which operate by lytic enzymes have already been outlined above (sections 1.3.3 and 1.3.4). Similar to these plate tests are those to check for siderophore production (section 1.3.2). Basically the antagonists are inoculated on a plate with the pathogen in much the same way as for an antibiotic test, but the medium is low in available iron; under these conditions siderophore producers should effectively compete for the limited iron and cause growth inhibition of the pathogen. If the test is repeated with added iron then the inhibition will disappear whereas in the case of an antibiotic the inhibition would remain, or possibly even be increased if excess iron encouraged the formation of secondary metabolites. Most siderophores that have been investigated are catechols or hydroxamates and there are simple colorimetric tests for the presence of these in culture filtrates, but caution is necessary, for the simple presence of the chemical does not tell you whether it is effective. Some pathogens produce siderophores that can out-compete the proposed antagonist siderophore for the iron. Tests for siderophore activity with the full system of host plant, pathogen and antagonist should also be done to see if this is still the mechanism. By using different chelating agents it is possible to make iron available in different degrees to the host, the pathogen or the antagonist (Swinburne, 1986).

General competition for space or nutrients is very difficult to test for easily, and as noted above (section 1.3.2) is not reported as the mechanism used by many control agents. In agar plate tests its detection will be very dependent on the medium used, and most laboratory media are so concentrated that nutrient shortage is unlikely in the short term. If there is inhibition in plate culture then it is difficult to prove that it is nutrient competition, rather than antibiotic production, except by showing that the inhibition disappears when excess general or particular nutrients are applied. The nutrient competition may be part of the general ability to show competitive colonization (section 1.3.1) and this can be tested for. It is possible to grow plants that are free from all micro-organisms by surface sterilizing seeds and growing 'clean' germlings in sterile conditions. The pathogen and the antagonists can then be added to the system to grow on the roots and leaves, as appropriate, and back isolation at various stages in growth will show how the colonization is proceeding and whether one is out-competing the other. This gnotobiotic system (containing only known organisms or mixtures of known organisms) is of course a greatly simplified system in that there is no confusion with all the rest of the organisms in a natural environment, and it is the first stage of colonization studies (section 1.3.1).

Another possible mode of action which must be checked is that of growth promotion (Fig. 2.3), but care is again necessary. The simple test is to grow the host plus the proposed control agent, in the absence of the test pathogen, and show increased growth of the host. In pots or in field tests this extra growth may however be caused not by actual growth promotion but by the control of minor pathogens or deleterious organisms which, though undetected, may be causing a loss of vigour (section 5.8). Tests on gnotobiotic plants will sort this out. Proper growth promotion or the control of minor pathogens is a perfectly valid means of biological control, which may allow the host to outgrow the main pathogen. The pathogen has not been controlled but a result that is potentially useful to the farmer has been produced.

Mycoparasitism is usually detected by lysis of colonies of the pathogen and may be confirmed by light or electron microscopy (Fig. 1.10).

We now have a disease, the symptoms of which can be reduced by our selected antagonists, that operate in known ways so that there is a suitable delivery system. It is now time for field trials. In practice the investigator rarely has the patience to wait for this and goes ahead with field trials while the investigations are still being made. Short-term research funding, the need to justify research expenditure by a practical demonstration and the need to produce a useable product ahead of commercial competitors, may not be the best determinants of a course of action. There is, however, the risk that potentially useful organisms will fail in field trials and be discarded when in fact they would work if only put in suitable numbers in the right place and at the right time.

2.7   Field testing potential control agents

There are various levels of field testing. It may mean: (1) Small experimental plots, perhaps only a few metres square, that are really used as an extension of the screening described above. (2) A specialized test, of any size, to look at colonization of the soil or the host, formulation of products, any effects of environmental factors, inoculum concentration, mixed inocula, cropping systems or other experimental variables that are thought to be important. (3) A full scale test of the proposed biocontrol agent under normal agricultural conditions. We will discuss these in more detail below, but in general field tests are very important in verifying that the organisms work outside the laboratory or glasshouse. They do, however, give very variable results due to climatic and soil differences from year to year and from place to place. It is, therefore, necessary to use large numbers of replicates of each treatment, for the high variance will make statistical proof of the results difficult. Field testing usually requires skilled personnel, often with special seed drills or harvesting equipment, to get yield data etc., from what are agriculturally very small areas, but nevertheless too large for the hand harvesting which can be used in laboratory experiments. Estimation of the amount of disease can also be laborious on a large scale. This leads to the taking of samples from the plots and using less rigorous assessments of performance than are possible on a small scale in the laboratory. Both these latter expedients increase the variability or decrease the precision of the data.

Field trials are time consuming, expensive and sometimes tedious but they are a very necessary part of the development of a biocontrol agent. The number of biocontrol agents that are pronounced excellent in the laboratory, and which result in scientific papers, is very large, but the number that are useful in commercial agriculture is very small. The difference between the two figures is the result of testing under field conditions.

The small plot trial as an extension of screening is just doing the work under very difficult, though more realistic, conditions. Usually there are too many organisms to test in the field, so laboratory screening is used, for better or worse, to reduce them to a number that can be handled.

The specialized experimental field test of the micro-organisms or formulations is usually a necessary preliminary to larger trials. In contrast to the laboratory tests with gnotobiotic systems, you are forced to consider the real situation in the field where there are lots of other organisms. The main problem with these experiments is to know that your test organism is still part of the experiment, that it has survived, grown, moved or colonized. You have to be able to find your organism, not just organisms of the same genus, species or biotype but of the same strain (Andrews, 1987). There may easily be one or ten million microbes per gram of soil or plant material and it may be necessary to find at least some of the 1000 or so of the required strain: you are looking for an individual in 1000 or 10000 others. There are media with varying degrees of selectivity for many genera, and pasteurization or incubation conditions can sort out many more. So it may simply (!) be necessary to distinguish between species or strains of a few genera that are isolated. With luck, or forethought, the required strain may be morphologically characteristic. The required organism may have some intrinsic resistance to toxins (such as heavy metals or antibiotics) which can be used to improve the selectivity of media. More important is that such tolerance can be induced or selected for, so that the biocontrol strain has resistance to unusual amounts or types of antibiotics and such a strain can then be isolated on media that exclude almost all other microbes in the environment to be searched. Again it is possible that in selecting for antibiotic resistance the biocontrol properties may be lost, and such selected organisms often grow more slowly than the wild type. Care is therefore necessary in interpreting the results with these antibiotic resistant strains. There is another way of finding your organism, and that is by using immunological methods to recognize it (Bohlool &

Schmidt, 1980). Essentially you inject your organism, or part of it, into an animal and later collect the antibody that the animal produces. After purification this antibody may specifically attach to the strain originally injected, though specificity varies a great deal. The antibody, now attached to the micro-organism, can then be located by specific binding of dyes or fluorescent stains. The sensitivity of such methods is not usually high and at present only numbers greater than about 10000 per gram are likely to be detected. There may also be cross-reactions in the soil with other organisms or organic materials. However, it is likely that these immunological methods will eventually be the method of choice, though much development remains to be done. This is a case of conflict in trying to stay within known technology to improve chances of success but at the same time needing to develop methods, which is a full-time research project in itself.

Finally, there is the ultimate test of the organism in agricultural conditions and on an agricultural scale, with no helping hands from specialist microbiologists, formulation chemists, field trial teams and so on. The proposed product, or perhaps a number of different versions of it, is given to the farmer who uses standard machinery and agricultural practice to see if it works. There will be repeats in different years, on different soils, with different crop varieties in different countries. At this stage there will already have been toxicology clearance and safety tests before the micro-organism is allowed to be released, and any commercial interests will have protected their research investment with some form of patenting.

2.8   Patenting biocontrol agents

If a biological control agent has been through the series of tests and screening outlined above, then a great deal of money has been spent and it is likely that the resulting product will have to be protected in some way from other possible producers so that the research and development money can be recovered from sales. In the case of a commercial company a profit will also be expected, and this is increasingly true also of universities and research stations who are having to fund research from the commercial exploitation of their expertise. There are quite a lot of different aspects of a biological control agent that may need this protection. There is the organism itself, the means of producing it in industrial quantities, the formulation of the product and possibly its method of use in conjunction with other forms of control such as fungicides or particular agricultural practices. Another, less tangible, item is 'intellectual property and know how' which is the knowledge and techniques involved in the research, development and use of the biocontrol agent.

The simplest form of protection, and the least effective, is industrial secrecy. During the development work the researchers would not be allowed to publish or talk about their work, and this would prevent detailed knowledge from reaching others, though inevitably people hear rumours that a researcher or company is working on this or that. Once the product is marketed the secret is out, and everyone can find out the organism and the formulation, if not the production methods and know how, by simply buying the product.

Much more reliable, though more expensive, is to try to establish some form of patent on the product. The patent protection offered, and the means of obtaining it, vary from country to country but there are international conventions and some of the European countries have established a common system (Crespi, 1985). Once having obtained a patent then you have a monopoly for 20 years, though you may license others to produce or market your invention. To obtain a patent it has to be shown that the product or invention has novelty, that it has not been previously used, talked about or published in any way which would allow anyone to gain a knowledge of it. There must also be an inventive step, something that the inventor has done to create the new product: you cannot patent a theory, an idea or anything you have found or come across that was there before but was just not noticed. Finally the object or process must have an industrial, including agricultural, use.

The most usual thing that is patented is the production process, cultivation techniques for the microbe and the particular formulation of the product. It is possible to patent a substance (e.g. an antibiotic produced by your organism) and to patent the micro-organism itself; the patenting of higher plants and animals is just being attempted but they may already be protected by systems of breeders rights (Crespi, 1985; Ruffles, 1986). There are, however, problems with patenting a micro-organism for it has to be described exactly, and as mentioned above there are problems with naming and identifying microbes from natural environments. The alternative is to describe it by saying that it is the same as culture number such and such in a recognized national culture collection. The disadvantages are that such a culture may change with time, and as there is public access to the culture, another person or company can obtain and study your isolate. Furthermore with a naturally occurring organism it is possible, once they know the sort of organism involved, to isolate their own strain from the wild which is just a little different from yours. You could make your organism special by altering it in some way, including genetic engineering, but then there would be severe problems at the present time with getting permission to release it into the environment (section 2.6). So patent protection is generally sought on the whole process and product, including the micro-organism, rather than on the microbe alone.

2.9   The future of biological control

In this chapter we have seen that there has been a rather long theoretical and academic history, with a recent phase of commercial interest. There is no doubt that there is a future, for there is a demand for something other than chemical methods of controlling plant disease (section 2.2) and there are also some successful biocontrol agents on the market. However, the lead-in time from the present commercial research is five to ten years, starting in the early 1980s. Furthermore the first biocontrol agents that are marketed will, in retrospect, probably appear to be very ineffective despite the research that will have gone into their development. Consider the early sulphur and copper fungicides of 100 years ago compared with the sophisticated modern chemicals that have had intensive development, costing millions of pounds or dollars.

It is necessary to consider the future of the intensive, mechanized agriculture dependent on artificial fertilizers and pesticides, separately from that of the subsistence, labour intensive agriculture of the Third World. It is possible in the long term that intensive western agriculture, which is producing food surpluses, may be replaced by more extensive, sustainable systems. Despite the basic desirability of sustainable systems, the present experience with overproduction suggests that the solution which will be adopted is to keep intensive agriculture to a smaller area, rather than to have systems with a lower input over a larger area.

In the less intensive systems the need is for inocula that, in combination with organic amendments and sustainable agricultural systems designed to maintain and encourage the control, will not need to be supplied continuously. Such a one-off sale means that the market is not commercially attractive and the commercial companies will not therefore produce such products. The research will have to be done by research institutes and universities, in the countries concerned, on a non-profitmaking basis. This sort of operation is similar to the development of long-term *Rhizobium* and mycorrhizal inoculants which, once in the soil in the presence of suitable hosts, do not need regularly replacing. So in the Third World we are talking about biological control in its widest sense, possibly combined with some inoculum of particular micro-organisms as starters. Much of this technology is developed locally and does not form a prominent part of the scientific literature, so it will not feature much in the following chapters. This unfortunate state of affairs reflects an inability on our part to discover what is going on, rather than a comment on the value of such systems.

In intensive agriculture there is a need, and a capability, to handle more sophisticated, high technology products. These are expensive, and probably designed to last rather a short time. If they die out quite quickly then they are environmentally more acceptable for they are less likely to spread to places where they are not supposed to be. This is also commercially attractive. Consider the replacement of persistent pesticides by the modern biodegradable ones which was done for good environmental reasons: there are also good commercial reasons, for if the pesticide disappears within a few weeks then the grower has to put it on again, and therefore buy it again, and this keeps the production going rather than having a virtually one-off sale.

It must however be understood that western agriculture is most unlikely to go to completely organic systems in the near future. The commercial, financial and agricultural dependence on fertilizers and pesticides is too great to allow any rapid change to sustainable agriculture. Future research on 'organic farming' may improve the situation, but for the present the biocontrol agents will have to work alongside pesticides and other agrochemicals in integrated control systems dependent partly on cultural techniques, partly on chemicals where safe, effective and environmentally acceptable ones exist, and partly on microbial inocula. The latter may be fungicide resistant and designed for particular cropping systems.

It is clear that most of the currently available biocontrol agents, and those next on the market for use in intensive agriculture, are in the horticultural sector where they are relatively easy to produce (section 2.3). However, there are major programmes for diseases, especially of the roots, of such main crops as potatoes, cereals of many sorts and legumes like soybean.

2.10   Conclusion

These first two chapters have, hopefully, set the scene for the detailed examples that now follow. We will tend to get involved in the minutiae of particular diseases and control systems, but do try to bear in mind the ecological principles and difficulties, and the problems with commercially orientated research in a subject where the methodology and active ingredients both fall short of the precision to which we have become accustomed thanks to the recent dominance of agricultural and analytical chemists. We are dealing with a dynamic, not to say totally unpredictable and unstable, system which we are trying to influence and control by very crude means. The resultant uncertain and conflicting results do not necessarily mean that the system itself is useless, just that we do not yet understand enough about it.

3

Biocontrol on leaf surfaces

3.1   Structure and microbiology of leaf surfaces

There are now many texts dealing with the microbiology of the phylloplane (the leaf surface) especially those based on a series of conferences which started in 1971 (Preece & Dickinson, 1971; Dickinson & Preece, 1976; Blakeman, 1981; Fokkema & van den Heuvel, 1987). The subject is also covered in many general texts on microbial ecology (Campbell, 1985) and Windels and Lindow (1985) have recently produced a small specialist study on biocontrol on the phylloplane that also contains some introductory material. The reader is referred to these texts for a general treatment of the microbiology of the phylloplane, and only an outline will be given here.

The surfaces of leaves are usually hydrophobic due to the presence of cutin and wax, the quantities of which vary with the plant species or cultivar and with the environmental conditions. This impervious layer not only restricts the loss of water from the leaf but also reduces the amount of nutrients which are leached from the leaf (cf. the root, section 5.1). Nutrients may also arrive on the leaf from dust and most importantly from the deposit of pollens. Very few leaves have flat or smooth surfaces: there are often crystals of wax of various shapes and the epidermal cells have convex surfaces with channels between them (Fig. 3.1). There may also be microscopic hairs even on apparently glabrous leaves. These topographical details give many microhabitats with improved water availability or nutrients or with protection from excessive radiation, the ultra-violet component of which is particularly damaging. Xerophytic plants may protect themselves from too much water loss by the growth of dense hairs or by rolling up the leaf to reduce the exposed surface and this of course protects the microbes as well.

The implication of these conditions is that the growth of micro-organisms on leaves is normally severely restricted by environmental factors, and any slight benefit that can be obtained from shelter is advantageous. Nutrient limitation is general, but the other factors vary with climate. Under temperate conditions, and of course in the arid tropics, water is frequently limiting and growth may only occur in rain or

Fig. 3.1. Micro-organisms on the leaf surface: (*a*) yeast cells concentrated in depressions between epidermal cells. (*b*) A rust spore (*Uromyces vicia-fabae*) on a leaf surface with many yeast cells. This is a very dense flora for a temperate leaf surface. (Photograph courtesy of Dr A. Beckett, Department of Botany, University of Bristol.)

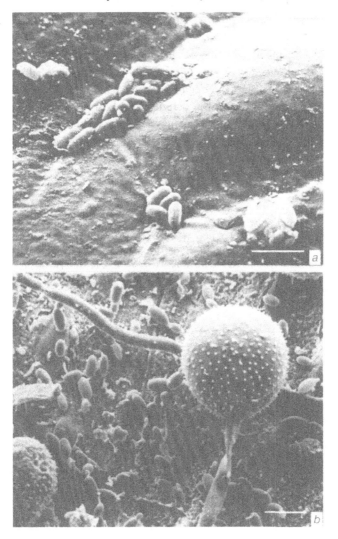

periods of dew, which can create free water or at least high humidity (Fig. 3.2). Despite what many of us who live in temperate oceanic climates may think, there are many spells of dry conditions on leaves. Conversely there may, even in such climates, be excessive irradiation for the leaf is designed to absorb radiation and its temperature may rise to several degrees above ambient. In the humid tropics, the ultimate expression of which is the tropical rain forest, the situation is very different with many leaves permanently wet. Under these conditions there can be very luxuriant growth with extensive microbial films.

So what microbes do grow as saprotrophs, especially on crop plants? In temperate conditions the most frequently reported organisms are the fungi *Aureobasidium pullulans*, *Cladosporium* spp. and yeasts such as *Cryptococcus* and *Sporobolomyces*. The few reports of accurately identified bacteria suggest that *Pseudomonas*, *Xanthomonas*, *Chromobacterium* and *Klebsiella* are usually present. These lists should however be viewed with some caution because they are based on cultural studies and so give no indication of the activity of the organisms: *A. pullulans* for example is known to be dormant for much of the summer period, even though it can be easily isolated. There are other seasonal changes, and different populations on the top and underside of leaves and on

Fig. 3.2. (*a*) The effect of different relative humidities on the development of *Sporobolomyces*. 65% RH (O——O); 75% RH (△—△); 85% RH (⊙—⊙); 95% RH (□—□). (*b*) Effects of dew (▽—▽); 95% RH(□—□); 65% RH (O—O); 65% RH to dew (O--- ▽); 65% RH to 95% RH (O---□). (From Bashi, E. & Fokkema, N. J. (1977). *Transactions of the British Mycological Society* **68**, 17–25.)

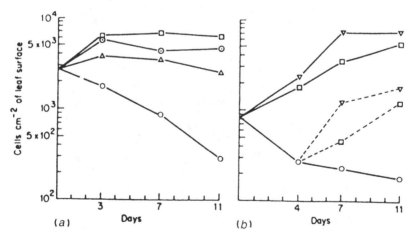

different parts of a tree canopy depending on the amounts of exposure in relation to the prevailing wind and rain.

3.2   Important leaf diseases

Important in this context means, unfortunately, those diseases which cause damage to economically important crops and, furthermore, damage of such an extent that the losses involved justify research on, and ultimately deployment of, expensive control methods. There are two outstandingly important groups of diseases, the biotrophic rusts and the mildews which have the potential, fortunately not often realized, for the destruction of major crops. The primary control measures at present are the extensive use of cultivars resistant, usually to particular races, of the pathogen. In the rusts, especially, this has now become very complex; particular cultivars of wheat, for example, being resistant to a limited selection of races of pathogens such as *Puccinia graminis, P. striiformis* as well as the mildew *Erysiphe graminis*. It is, therefore, necessary to know the races of particular pathogens that are in the district before selecting the cultivar to be grown, though there will be other considerations, such as those connected with the agronomy and market forces. There is also an extensive use of fungicides in complex spray programmes to control actual or expected outbreaks of diseases which are forecast by epidemiological studies. Even with these generalities it is important not to underestimate the differences in the basic biology of the rusts and mildews which greatly affect the control strategies. Rusts germinate and rapidly penetrate the leaf, having only a short period when they are vulnerable to biocontrol by surface inhabiting microbes: they cannot be 'reached' again until they emerge from within the leaf to release their spores (sections 1.2 and 3.6.3). In contrast to this the mildews are almost entirely confined to the leaf surface, with only haustoria in the epidermal cells. They are, therefore, accessible to control agents for almost their entire life history, though such organisms would have to be as well, or better, adapted to the stressed leaf surface habitat as the mildew itself.

Apart from the cereals mentioned above there are so many other rust and mildew diseases that it is difficult to choose representative examples. There is *Melampsora lini* on flax, coffee rust (*Hemileia vastatrix*) and rusts of flowers and ornamental plants. There are mildews on most commercially exploited plants from specialist crops like hops (with *Pseudoperonospora humuli*) to the many serious infections of cucurbits (*P. cubensis, Sphaerotheca fuliginea*) in glasshouse culture and in the field.

The next group of major leaf diseases are the necrotic leaf spots that range from the many trivial disfigurations of leaves to a few major problems which are mostly controlled at present by host plant resistance and/or spray programmes. The main importance of *Venturia* is in the formation of scabs on the apple fruit, but the overwintering stage is on the leaf on the ground as a leaf spot. Similarly, though *Phytophthora infestans* causes a leaf disease of potato its main effect is on the yield of tubers and their subsequent decay in storage. Many of the 30 or so leaf spots listed as major world diseases (Johnston & Booth, 1983) cause yield losses. Leaf spots are potentially capable of control by biological means, but usually there are many suitable fungicides already available, and anyway many of them are of minor importance.

Finally, in this brief consideration of the main groups of leaf diseases there are those caused by plant viruses which may result in various forms of chlorosis and discoloration, in the mosaic, streak and mottle diseases, and in growth distortions. There is some exploitable host resistance which may be strengthened by induced resistance (sections 1.5 and 3.6.2.) produced biologically. There are no chemical controls for virus diseases though sprays to control vectors such as aphids may be used.

3.3   Effects of fungicides on non-target organisms

The use of fungicides, which affect micro-organisms other than the pathogen at which they are directed, may well give a first clue to the existence of a natural control action by the resident saprotrophs. Fungicide effects apply to all parts of the plant, but since by far the majority of fungicides are used against leaf diseases we will discuss it here (though see section 5.9 for cereal stem base effects). Some fungicides are general toxins and have a very wide spectrum of activity: this is true of the older mercury, copper and sulphur formulations and some of the more recent protectant and systemic compounds based on synthetic organic chemicals (e.g. some benzimidazoles, captafol or dithiocarbamates). There clearly has to be some differential toxicity between the fungi and the host so that there is the minimum of phytotoxic effects. There are now fungicides that have differential toxicity for different fungi, as well as the difference between host and pathogen: such fungicides may be genus specific like tridemorph or ethirimol used against cereal mildew (*Erysiphe*) and they have only minor effects on a few other fungi. A further complication is that many spray programmes involve the use of several fungicides, which may even be mixed in the same tank, not to mention insecticides and other pest control agents. It is, therefore, very difficult to estimate what effect pesticides may have on the resident microflora or on antagonists introduced onto leaves.

Many pathogens, and some saprotrophs, can and do develop resistance to chemical control agents and the more specific a fungicide is the greater the likelihood of resistance developing. General toxins that inhibit a lot of different enzymes are difficult to develop resistance to, but compounds like benomyl (a benzimidazole) which specifically affect the production of a protein, tubulin, has now become much less useful because of the development of widespread resistance by such fungi as *Botrytis*. This is in addition to the fact that, though benomyl has quite a wide spectrum of activity, there have always been groups of fungi, and even some particular genera, that are not affected (e.g. *Rhizoctonia*, *Alternaria*, *Helminthosporium* and most phycomycetes). So a spray of benomyl may kill some strains of *Botrytis* but leave others to grow, kill some saprophytes but not *Alternaria* and leave some pathogens unaffected: this clearly upsets the normal balance of microbes on the leaf. Many of these saprotrophs antagonize pathogens by nutrient competition so their removal may make the disease worse.

Despite all these well-known generalities there are remarkably few careful studies published on the effects of fungicides on non-target organisms. There are many anecdotal stories and asides in various studies where the reduction of this or that organism was noticed. Alternatively there may be increases in diseases other than that whose control was being attempted, which may indicate that a pathogen previously controlled by natural means has become important because of the altered balance. However, Fokkema & de Nooij (1981) did study, in culture and on leaves, the effects of various fungicides on leaf surface saprophytes that have been used as biocontrol agents. Firstly Table 3.1 shows the *in vitro* sensitivity to fungicides of the potential control agents: notice that the wide spectrum fungicides at the top of the list allow almost no growth of the saprotrophs, so the application of these chemicals would destroy any natural control that was occurring. This is confirmed in the field where just one treatment can almost eliminate the population, though it later recovers (Fig. 3.3). Fungicides towards the bottom of the list in Table 3.1 have less effect on the saprophytes, while still controlling the target pathogens. Use of such chemicals, therefore, allows advantage to be taken of any natural control as well as the chemical control provided by the fungicide. Fig. 3.3 also shows the effects of mixed fungicides; one treatment with triadimefon would probably have had little effect judging by Table 3.1, but the combina-

Table 3.1 *Sensitivity of the saprotrophs* Sporobolomyces roseus *(SPOR)*, Cryptococcus laurentii *var.* flavescens *(CRYPT)*, Aureobasidium pullulans *(AUR) and* Cladosporium cladosporioides *(CLAD) to various fungicides on potato dextrose agar*

Fungicide	conc. g/litre[a]	SPOR	CRYPT	AUR	CLAD
Captafol[b]	0.10	∅	−	−	−
Captan[b]	0.12	∅	∅	−	−
Mancozeb	0.48	∅	∅	∅	∅
Maneb	0.48	∅	∅	∅	∅
Thiram	0.08	−	−	±	±
Benomyl[b]	0.05	∅	+	∅	∅
Carbendazim[b]	0.05	∅	+	∅	∅
Thiophanate-methyl[b]	0.14	+	+	∅	∅
Tridemorph[b]	0.11	∅	∅	−	±
Prochloraz	0.08	+	±	∅	−
Triadimefon	0.03	+	+	±	±
Triforine	0.04	+	+	+	±
Ethirimol	0.06	+	+	±	+
Oxycarboxin	0.08	+	+	+	+
Sulphur	0.8	+	+	+	+

[a]Active ingredient, 10% of the concentration recommended in sprays.
[b]Also tested on leaves with saprophytes. Similar results were obtained except that the benzimidazole fungicides suppressed *Cryptococcus* and *Sporobolomyces*. Also the other fungicides, not tested *in vivo*, may behave differently on leaves.
Colony size with the fungicide has been expressed as a percentage of that without the fungicide (100%) using the symbol ∅ = no growth; − = <40% growth; ± = 40–60% growth; + = >60% growth.
From Fokkema & de Nooij, 1981.

tion with wide spectrum carbendazim and maneb proves lethal to all but some *Sporobolomyces*, though again the populations recover quite quickly. Multiple applications have more drastic effects and recovery takes longer. *Aureobasidium* is apparently sensitive to all the fungicide combinations and never recovers. These authors have shown that *Septoria nodorum* and *Cochliobolus sativus*, both necrotrophic pathogens of leaves, can be reduced by artificial inoculation with *Sporobolomyces*. The application of unsuitable fungicides or combinations could therefore reduce the biological control capacity of the saprotrophic microflora. Fig. 3.4 shows how the antagonistic capacity has been challenged by inoculation with the pathogen in the presence of a wide spectrum fungicide, benomyl. Sprayed wheat had fewer saprotrophs per

Fig. 3.3. Effects of fungicides on the occurrence of saprotrophic fungi on the flag leaves of wheat in the field. P = pink yeasts, mostly *Sporobolomyces*; W = white yeasts, mostly *Cryptococcus*; C = *Cladosporium*; A = *Aureobasidium*. Decimal growth stages of the wheat are show in ( ) after the sampling date. The one treatment fields received on 25th June 0.3 kg Bayleton (triadimefon) + 2 kg Bavistin M (carbendazim/maneb) per ha. The five treatment fields had in addition on 5th July 5 kg Goldion (sulphur/mancozeb) per ha and three treatments with Benlate, maneb and Bayleton on 12th, 25th May and 5th June. (From Fokkema and de Nooij, 1981. See Table 3.1.)

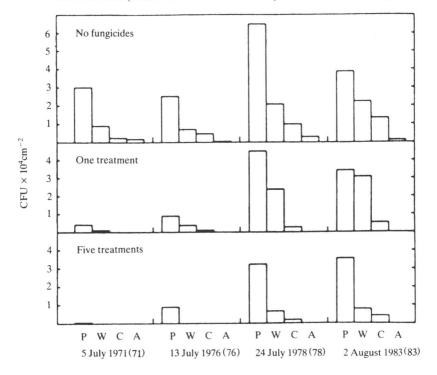

unit area of leaf than the unsprayed control. *Cochliobolus* was inoculated at the three dates and the benomyl treated leaves, with the lowest microflora, developed up to twice as much necrotic leaf area as the control with more saprotrophs. There was a good inverse correlation between antagonist density and the amount of disease. Studies on the saprotrophs showed that the natural *Cladosporium* and *Sporobolomyces* had been much reduced, and *Aureobasidium* eliminated by the benomyl: only *Cryptococcus* survived. These results agree with the predictions in Table 3.1.

74   *Leaf surfaces*

Other authors have reported similar effects from the field. The use of benomyl has been associated with increases in *Cochliobolus* and *Helminthosporium* leaf spots of wheat and this or related fungicides may be responsible for the recent increase in sharp eyespot (caused by *Rhizoctonia* which we noted above was insensitive to benomyl). If the balance of pathogens and saprotrophs is disturbed by the selective removal of particular components of the microflora then previously unimportant pathogens may be selected for their ability to survive the stress imposed by the fungicide. Secondary diseases like those caused by *Alternaria* and *Cladosporium* may form sooty deposits on leaves, especially in the presence of honeydew or pollen and they are encouraged by benzimidazoles to which they are insensitive. The growth of *Alternaria* on cauliflowers, previously an unimportant problem, is also made worse by the use of some fungicides against other diseases.

Fig. 3.4. Seasonal development of the total mycoflora of rye leaves sprayed with benomyl (+B, ●, hatched columns) or water (−B, ○, open columns) and its effect on the infection by *Cochliobolous sativus* in 1973. The dots show the total number of saprotrophic propagules on a given date. Columns, for each inoculation date, show the necrotic leaf area caused by *C. sativus*. The spray reduces leaf saprophytes and increases disease. (From Fokkema *et al.*, 1975.)

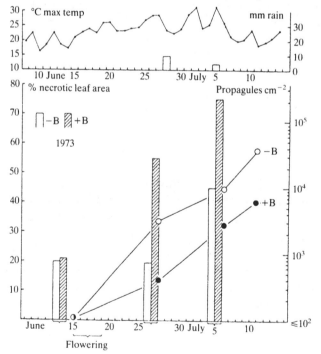

Leaf rust of coffee (caused by *Hemileia vastatrix*) is not normally a serious disease. It occurs sporadically in August and September in East Africa (control lines on Fig. 3.5). Fungicides are used on coffee for disease control and for yield increases independent of disease, which

Fig. 3.5. Effects of different fungicides on rust (*Hemileia vastatrix*) on coffee leaves. □ — □ = unsprayed control; ○ — ○ = sprayed in 1969 only; ● — ● sprayed in 1969 and 1970. A single spray gives more rust than no spray. (From Mulinge & Griffiths, 1974.)

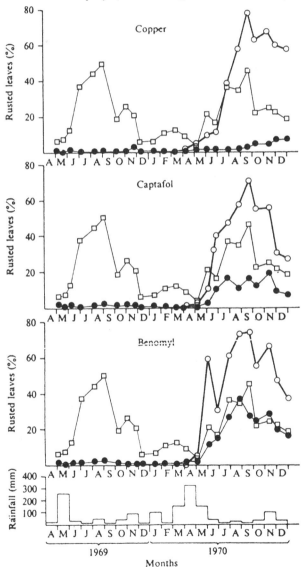

may be related to delayed leaf fall (possibly itself a microbiological effect). Such fungicides will also control coffee rust if applied at just the right time at the beginning of the rainy season, but if the spray is wrongly timed, not used or otherwise ineffective then the disease is worse on those bushes that were sprayed the year before than on those where control has never been attempted (Fig. 3.5). Copper in trace amounts, from fungicide residues for example, increases the germination of the rust uredospores, but it is also likely that the reduction in natural saprotrophs by the previous sprays removes this source of control and allows the disease to be worse, unless the current fungicide application is completely effective. There may be no advantage in attempting to control the rust with benomyl, it works the first year when there are saprotrophs, but subsequent applications do not reduce disease below the control levels and the disease may indeed be worse than before if spraying is stopped.

These sorts of reactions represent the destruction of existing, natural biocontrol which has been reducing disease levels without intervention by man, and is indeed only revealed by such intervention. It is, therefore, important to preserve and use this balance by careful restriction of the choice of chemical controls. This natural activity can be enhanced by artificial inoculation of native or exotic antagonists, the mode of action of which we must now consider.

3.4   Modes of action and problems of biocontrol on leaves

The potential modes of action (section 1.3) are limited by the harsh environment on many leaf surfaces. In temperate conditions competition for space is unlikely (Cullen & Andrews, 1984) for the amount of the leaf surface covered is usually much less than 1% and most of the pathogens anyway rapidly disappear inside the leaf, leaving the potential saprotroph competitors on the surface. Despite this some of the fungicide results discussed above do suggest that competition may occur, possibly in microhabitats limited for protected space as opposed to competition for the total available space. The only place where competition for total space may occur is when leaves reach the litter and biocontrol may be possible by competition at this stage to reduce overwintering inoculum (see *Venturia*, section 3.6.1). In the tropics, especially the wet humid regions, there may be competition in the continuous microbial films (section 3.1), but so little is known about such communities that it is difficult to produce any actual data to support this. The reports about such films, which are widespread in the literature, are almost all based on a single pioneering study by Ruinen (1961).

Nutrient competition has been shown for many situations and will be discussed more extensively below. Such systems as have been investigated are not concerned with the long-term effects and combative *K*-strategists (section 1.3.2) but rather with *r*-strategists taking advantage of temporarily improved conditions, during which their rapid germination and high growth rates allows them to exploit the situation, sometimes to the detriment of the pathogen.

The ability of many leaf surface organisms (e.g. *Aureobasidium, Sporobolomyces*) to produce antibiotics against both fungi and bacteria is not in doubt, when they are tested in culture. There is, however, no information on the production on leaves. Indeed the few times that organisms selected for antibiotic production have been used, there has been very limited success.

There has been much work on mycoparasites on leaves, especially with various rust fungi, and this may be a long-term measure to reduce inoculum (section 1.3.4).

The biochemical and morphological defence mechanisms of the host (Bailey & Deverall, 1983) have been studied on the aerial surfaces of plants and it has been suggested that potential antagonists may operate by stimulating the defences in advance of the pathogen.

So because the leaf surface is a highly stressed environment, often with only intermittent periods which allow growth in temperate conditions, it is a very difficult environment into which to introduce antagonists and have them survive and multiply. In most cases the introduced organisms die very quickly and are not maintained in numbers sufficient to be effective. There are many potential biological control agents for use on leaves, which have been described in the literature, but none are available commercially. This may reflect the above problems, especially if the antagonists are selected in the wet, high nutrient conditions of agar plates. Where control has been demonstrated in the field it has usually involved the addition of nutrients and/or the maintenance of high humidity and possibly free water on the leaf to remove these environmental restraints. The disadvantage of such treatments is that, apart from cost, they may also favour the growth of the pathogen! It is also found (Cullen & Andrews, 1984) that organisms isolated from environments other than the leaf (*Pseudomonas cepacia, Trichoderma viride, T. pseudokoningii* and *Myrothecium*) may show promise. This suggests that perhaps the prime consideration, especially for specialized pathogens, is to find fast growing, dark spored organisms regardless of where they usually grow in the field. The surface growth of pathogens like *Alternaria, Botrytis* and *Colletotrichum* may be more vulnerable to nutrient competition from control agents.

This all suggests that the outlook for biocontrol on the leaf may be different from that being developed for soil systems where most work has been done. For leaves it may be necessary to look for short-term treatments, not expecting or hoping for the long-term survival of the introduced organisms or the establishment of a new ecological balance to the disadvantage of the pathogen.

3.5   Germination inhibition and lysis

Germination of fungal spores on the leaf surface is a critical stage in the development of the host–pathogen interface, and one in which the pathogen is often very vulnerable. The physiology of germinating spores has been much studied, though admittedly mostly *in vitro*. All spores require water to imbibe, swell and germinate, and there are some larger spores for which this is the only requirement since they have sufficient endogenous food reserves for their initial growth stages. Other spores require an exogenous supply of sugars and sometimes amino acids. The presence of nutrients in the water can break spore dormancy and hence the general fungistasis of many environments (Lockwood & Filonow, 1981). Clearly those spores that require exogenous nutrients may be subject to competition from the saprophytes for the available nutrients including those from pollen and aphid honeydew. Even those spores that have their own nutrients may suffer competition because as they break dormancy and become hydrated the plasma membrane is temporarily disorganized and leaks organic materials. Normally this material is rapidly re-absorbed by the spore, but in the presence of saprotrophs this nutrient may be used by others, preventing the germination or the growth of the germ tube and leading to lysis of the pathogen spore.

Such a sequence of events has been studied in detail for the conidia of *Botrytis* and has been referred to before (Fig. 1.4). The germination of *Botrytis* spores is inhibited on the leaf surface in the presence of bacteria and yeasts, and the degree of inhibition varies with different isolates of bacteria. Inhibition is reduced or eliminated by the supply of exogenous nutrients, especially sugars and amino acids, and by removing the bacteria. Further studies with *Pseudomonas* and other microbes showed that they took up ^{14}C label supplied in amino acids and that the amount of uptake was related to the amount of germination inhibition (Fig. 3.6). High humidities are required for this activity, as the water drop must be maintained around the spore and the control of *Botrytis* by this means has not been shown under normal field conditions.

Similar results were obtained for *Sporobolomyces* antagonizing the pathogen *Meria* on larch needles: the reduction in spore germination caused by *Sporobolomyces* was reversed by the supply of nutrients, indicating that nutrient competition was the most likely mode of action. In the same studies, however, the germination reduction caused by *Pseudomonas* was shown to be unaffected by nutrients, which implies that some other mechanism is involved. In both situations the plants were maintained for the test at very high humidities and control was not shown under field conditions.

There is also some variation in the effects of different nutrients in some combinations of pathogen and potential antagonist. *Drechslera dictyoides* is a fungus that causes a leaf disease of perennial rye grass and the germination of its spores can be delayed by various bacteria which also reduce the amount of surface hyphal growth (Table 3.2). The cytoplasm of many of the hyphae is also destroyed or damaged (as judged by staining reactions) in the presence of the bacteria. All isolates of the antagonists reduced the amount of disease when inoculated onto

Fig. 3.6. Scatter diagram of the % germination by *Botrytis cinerea* conidia after 24 h in droplets containing bacteria or yeasts on leaves, against % uptake of amino acids. □ = *Pseudomonas* isolate 14; ○ = *Pseudomonas* isolate 9; ● = *Pseudomonas fluorescens* isolate 15; ▲ = isolate 11b; △ = *Sporobolomyces* isolate CH7; ■ = natural epiphytic population. The regression line obtained using similar organisms *in vitro* is also shown (-----). (From Blakeman, J. P. & Brodie, I. D. S. (1977). *Physiological Plant Pathology* **10**, 29–42.)

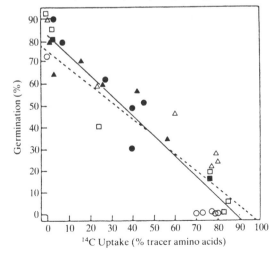

the plants, though the amount of disease control did vary (Table 3.2). The addition of glucose or yeast extract to the bacterial inoculum greatly increased the effectiveness of one *Pseudomonas* and *Xanthomonas*, but only the yeast extract improved the antagonism of another *Pseudomonas* and *Listeria* (Table 3.2). In these cases it seems that it is definitely not nutrient competition, indeed added nutrients are necessary for the bacterial antagonists though they have no effect on the *Drechslera*, which germinates satisfactorily in water alone. Despite the maintenance of high humidity the applied bacteria did not survive for more than a week or two on the leaves, the control occurring at germination and the early growth stages of the pathogen, was possibly by the production of antibiotics by the bacteria.

The artificial addition of nutrients in this study has some similarities

Table 3.2 *Effects of various antagonists on the germination and growth of* Drechslera *spores and mycelium*

Potential antagonists			*Listeria denitrificans*	*Pseudomonas fluorescens*	*Pseudomonas fluorescens*	*Xanthomonas campestris*	Bacteria-free control
Germination %		3	82	85	91	70	97
on tap water		6	91	96	96	91	100
agar at 3, 6, & 9 h		9	100	100	100	100	100
Germ tube length		3	50	46	48	28	52
at 3, 6 & 9 h		6	127	123	127	139	142
µm		9	255	255	203	238	278
	Inoculum	7	0	0	0	0	0
	&	14	0	0	0	0	0
% of first	water	21	13	7	27	73	85
leaves of	Inoculum	7	0	0	0	0	0
Lolium	&	14	13	20	0	0	20
infected	glucose	21	13	25	0	7	85
at 7, 14 &	Inoculum	7	0	0	0	0	0
21 days	&	14	0	0	0	0	20
	yeast	21	0	0	7	7	85

Selected from Austin, Dickinson & Goodfellow, 1977.

to pollen deposition on leaves which has been shown to have considerable effects on the saprophytes and the pathogens. Firstly, it may greatly increase the amount of disease and the growth on the surface of the leaves (Table 3.3). Addition of antagonists can reduce both the disease and the mycelial growth to the level before pollen stimulation, and the germination of the *Drechslera* was also reduced in some, but not all, of the *in vitro* experiments. These effects were considered to be due to competition for the nutrients available from the pollen.

The above examples illustrate the story of many biocontrol agents for leaves, which are based on germination inhibition. Mechanisms of action are known, potentially useful organisms exist, and laboratory studies (often in high humidity) show control of the disease but in the field it fails because of the environmental stress, especially dryness, on the leaf.

3.6   Control of leaf diseases after germination

There are three main subjects for discussion in this section, firstly, the control of numerous diseases by mechanisms other than germination inhibition, secondly, the phenomenon of induced resistance and thirdly, mycoparasitism.

Table 3.3 *Effects of various antagonist isolates on* Drechslera sorokiniana

	Mean necrotic leaf area %			Superficial mycelium $\mu m/mm^2$	
	Expt. 1	Expt. 2	Expt. 3	Expt. 1	Expt. 2
Drechslera sorokiniana (D.s.)	3**	1**	6***	100**	125**
D.s. + pollen	57	72	64	2700	1450
D.s. + pollen + Cladosporium 42	10**	36*	17**	—	—
D.s. + pollen + Cladosporium 33	36*	42*	6*	—	—
D.s. + pollen + Aureobasidium	—	—	—	125*	325*
D.s. + pollen + Cladosporium 50	—	—	—	50**	275**

*, **, *** indicate significant differences at $P = 0.05$, $P = 0.01$ and $P = 0.001$ from the pollen control.
Selected from Fokkema, 1973.

3.6.1   *General control*

The many reports of laboratory studies have been described in several reviews (e.g. Blakeman, 1981) so we will limit the discussion to a few examples where control has been shown under something approaching normal growing conditions. As long ago as 1951 there were reports of biocontrol of leaf diseases. Wood and Newhook both showed a decrease in disease on lettuce by the use of fungi and bacteria, especially when the organisms were used as protectants by spraying the plants before exposure to the pathogen such as *Botrytis* (Table 3.4). These experiments were conducted under protective glass frames where high humidity could be maintained, but this is similar to some commercial production systems for lettuce. However, Wood did consider that repeat inoculations of the potential antagonists would be necessary and that commercially available fungicides (in 1951) were likely to be more effective. Since then systemic fungicides, such as benomyl, have come into use and were very effective, but resistant strains of *Botrytis* have developed. Thirty-five years on and complex mixtures of fungicides are now used but we still have no very good single fungicide for use on lettuce (there are problems with toxic residues on salad crops), and *Botrytis* and *Rhizoctonia* are still a problem. Various antagonists, such as *Trichoderma*, are being looked at again and we may yet have biological control in this situation. It is favourable for biocontrol because humidity can to some extent be controlled under glasshouse conditions, which may alleviate the problem of drying and death of antagonists.

Table 3.4 *Control of rot of lettuce leaves (caused by* Botrytis*) by a variety of antagonists under protected conditions of growth*

Antagonist	Simultaneous inoculation		3 days pre-inoculation	
	Little or no rot (%)	Extensive rot (%)	Little or no rot (%)	Extensive rot (%)
Control	27	73	17	83
Bacillus dendroides	41	59	75	25
Pseudomonas sp. no. 1	75	25	90	10
Streptomyces lavendulae	20	80	60	40
Streptomyces sp. no. 1	71	29	85	15
Penicillium clavariaeforme	80	20	95	5
Triochoderma viride	50	50	75	25
Fusarium sp. no. 1	95	5	100	0

From Wood, 1951.

Leptosphaeria nodorum (= *Septoria nodorum*, glume blotch) and *Cochliobolus sativus*, which attacks leaves as well as roots and stem bases, are both important diseases of wheat. The work of Fokkema's group on the growth of yeasts on leaves has already been mentioned (sections 3.1 and 3.3) and some control of these diseases has been shown. Yeasts (*Cryptococcus* and *Sporobolomyces*) were sprayed onto the plants with a nutrient solution and they increased in numbers (Fig. 3.7 and Fig. 3.8) in relation to controls sprayed only with water. Both *Septoria* and *Cochliobolus* were reduced for some time after the application of the yeast and nutrients, but when the natural flora reached a certain density (about 50 000 cm^{-2}) then this gave natural biocontrol and the extra yeast from the direct application did not further inhibit the disease. So there can be control of diseases in the early stages of development by supplementing the natural antagonistic flora, and at the same time removing the nutrient limitation so as to allow growth.

Another species of *Cochliobolus* (*C. miyabeanus*), which causes a brown leaf spot of rice, has also been controlled by the use of yeasts and other micro-organisms sprayed onto leaves. The mode of action is not

Fig. 3.7. Seasonal development of the total saprotrophic mycoflora on wheat flag leaves sprayed (↑) with yeast and nutrients (+Y, ●—●, hatched bars) or with water (C, ○——○, open bars). Graph is of numbers and the bars are the amount of infection by *Septoria nodorum* (% reduction in chlorophyll *a* content). Significant differences, * $P = 0.05$; ** $P = 0.01$ between treatment and control. (From Fokkema *et al.*, 1979.)

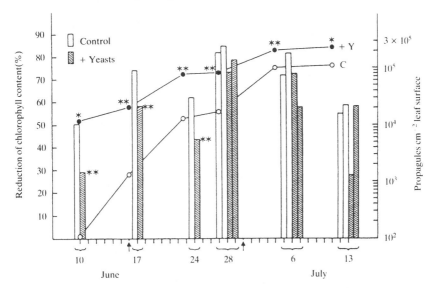

clear, for nutrients were not required and none of the isolates produced
inhibitory or lytic substances in culture.

Nutrients are, however, involved in a more complex way in the
control of apple scab (*Venturia*) during its overwintering stage. If urea is
sprayed onto the leaves just before leaf fall then it greatly encourages
the growth of saprotrophic fungi such as *Alternaria* and *Cladosporium*
on the overwintering leaves on the ground below the tree (Fig. 3.9) and
there is a reduction in the number of perithecia developed by the
Venturia and hence a reduction in the inoculum in the following year.
Apart from this microbiological effect there also seems to be a chemical
effect, with the nitrogen source preventing the maturation of the asci of
Venturia (Fig. 3.9). By using the overwintering stage of the pathogen,
with the leaves on the soil, the normal limitation of antagonists by lack
of water is removed and the nutrient-stimulated growth of the
antagonistic saprophytes can be utilized.

3.6.2 *Induced resistance and cross-protection*
This subject has been considered in general terms already (section 1.5)
and is a widespread phenomenon (Sequeira, 1983), but since much of
the work has been done on leaf diseases, especially on cucurbits (Kuć,
1981), it seems best to describe it here. *Colletotrichum* causes a number

Fig. 3.8. As Fig. 3.7, but the infection was with *Cochliobolus sativus*.

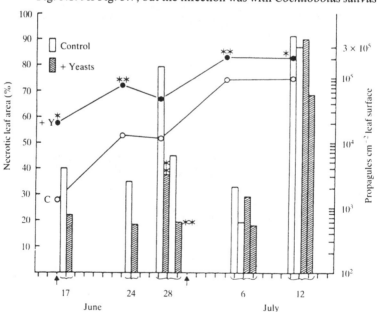

Fig. 3.9. Effects of urea spray applied on 17th October on the numbers of *Alternaria* spores (*a*) and *Cladosporium* spores (*b*) washed from overwintering apple leaves. (*c*) The stages of development of the perithecia of *Venturia inequalis* on these leaves with and without urea. (From Burchill, R. T. & Cook, R. T. A., 1971.)

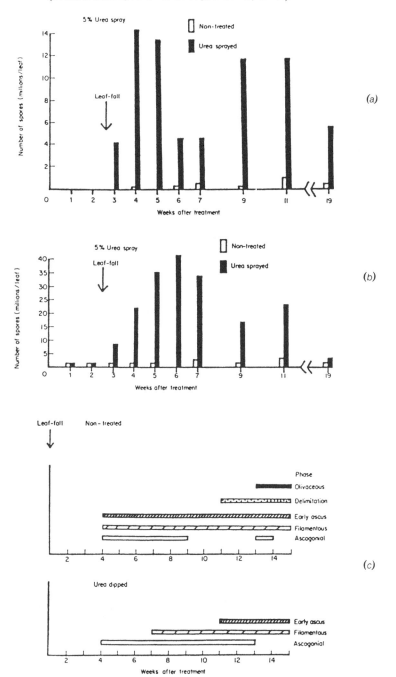

of anthracnose diseases and there are host-specific species and variation in the virulence of a species on a particular host. *C. lagenarium* causes cucumber anthracnose; *C. cucumerinum* causes scab in cucumbers; *C. lindemuthianum* causes anthracnose of beans. Cultivars of cucumber resistant to scab and inoculated with *C. cucumerinum* (though they do not develop the disease) become resistant to *C. lagenarium*. Inoculation with *C. lindemuthianum*, which never causes disease on cucumbers, makes the plants resistant to both *C. cucumerinum* and *C. lagenarium*. This is rather complicated, but basically what it says is that inoculation with an avirulent race or a non-pathogenic species gives protection against a pathogen, and furthermore the treatment can be applied to an early leaf and protection subsequently appears in leaves that are produced as the plant grows, even though the later leaves have not themselves been challenged. The resistance, or rather the presumed chemical elicitor of resistance, travels systemically in the plant even reaching leaves that were not expanded at the time of the initial challenge. The inoculated leaf can even be removed a few days after challenge and yet the induced resistance remains in the plant (Fig. 3.10).

Another variation on the technique is to challenge the first leaf on the plant with the real pathogen, wait until induced resistance appears in subsequent leaves and then remove the initial infected leaf, leaving a plant with no infection but with induced resistance. This is illustrated in Fig. 3.10, which shows that as long as the first leaf is left on the plant for

Fig. 3.10. Effect of time of excision of the inducer leaf (leaf 1) from the cucumber cultivar Marketer inoculated with *Colletotrichum lagenarium* (shaded bars) on the number of lesions on leaf 2 seven days later. Open bars are leaves inoculated with water only on the first occasion. Significant differences from the corresponding control at * $P = 0.05$; ** $P = 0.01$; *** $P = 0.001$. (From Dean & Kuć, 1986.)

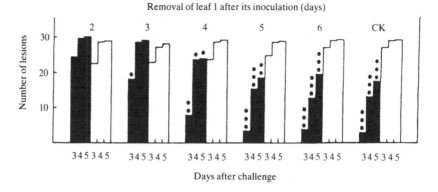

3 or 4 days then there is some benefit to subsequent leaves in terms of reduced numbers of lesions and also in reduced size of the lesions, as shown in Fig. 3.11, when the plant is subsequently challenged with the disease. There is a major problem with the system in that despite much work the translocated elicitor has not been identified or even isolated. Such a substance must be produced by, or in response to, many different fungi, bacteria and viruses as all can cause induced resistance, and it must have a rather low molecular weight for ease and rapidity of translocation. It is often assumed that this elicitor then activates the plant's defence mechanisms (section 1.5) including the production of phytoalexins and lignification of the walls, and that this is the induced resistance. Other possibilities include electrical signals, changes in

Fig. 3.11. Protection of leaf 2 against disease caused by *C. lagenarium* after inoculation of leaf 1, then excision of the tip (– – –) or of the whole leaf (——). Data are from 4 (●), 5 (■) and 6(∗) days after challenge of leaf 2. Control plants were inoculated with water on the first occasion. (From Dean & Kuć, 1986.)

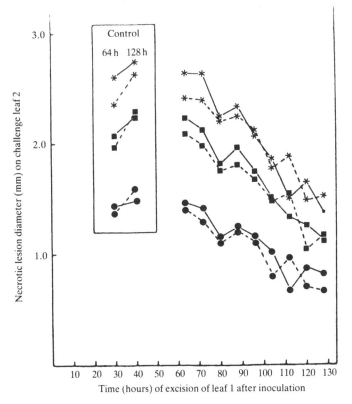

membrane potentials or ion fluxes as possible messengers, which would explain why no chemicals have been found. Control not shown to be via the plant's defence mechanism is referred to by the more general term of cross-protection.

Such a system allows the initial challenge to be carried out for a short time under favourable conditions of humidity etc., though the effect may last for a considerable time after the plant is put out in the field or the glasshouse. Notice in Fig. 3.10 however that, as usual with biocontrol agents, the disease is not eliminated. It may still require some chemical control, though hopefully less frequently or at a later stage in growth than without the biocontrol agent. This may be one of the reasons why the method is not in widespread commercial use; biocontrol is not enough on its own to reduce the disease to acceptable levels, so if you are going to use fungicides eventually it may be less trouble to the grower to use them all the time. However, further development of this and similar systems may lead to a more important role in the future.

3.6.3   *Pathogens of pathogens*

Fungi and bacteria causing diseases on leaves may themselves be attacked by pathogens. Most of the work has been done with those fungi that attack other fungi (the mycoparasites or hyperparasites, Kranz, 1981), but *Bdellovibrio* has also been used on leaves against bacterial pathogens. *B. bacteriovorus* is a widespread bacterium that can attack other bacteria, especially gram negative ones, by attaching itself to the outside, penetrating the cell and causing lysis, subsequently growing and dividing inside the 'host' bacterium. Different strains of *B. bacteriovorus* differ in their virulence to *Pseudomonas glycinae*, the cause of blight of soybeans (Fig. 3.12); in this case the strain Bd–17 is most effective at controlling both the systemic symptoms and the local lesions. Bd–19 has the same effect but Bd–10 is not virulent on the *Pseudomonas*. Increasing the proportion of *B. bacteriovorus* in relation to the *Pseudomonas* increases the degree of control of the disease (Fig. 3.12). The control will have the usual problems with lack of water on the leaves.

Much more widely studied in relation to biocontrol are the mycoparasites. There are about 210 species known, including those that operate in the soil like some *Trichoderma* strains, and there are 84 species that parasitize rust and powdery mildews (Kranz, 1981). Out of this long list of reports there are in fact only 4 species that have been studied in much detail and these are: *Sphaerellopsis filum* (mostly as its anamorph *Darluca filum*) which parasitizes uredia and telia of some 362 species of rusts, *Tubercularia vinosa* on pycnia and aecia, and *Verticillum lecanii* (also an insect pathogen) on uredia. *Ampelomyces quisqualis* parasitizes all stages of powdery mildews. The mycoparasite usually penetrates the host hypha or spore and kills it. It seems, however, that some of the control may be by physical displacement of the rust or mildew, overgrowing the sporing pustules and preventing spore dissemination even if the spores themselves are not killed.

The problem with mycoparasites is that, despite their world wide distribution and their common occurrence, they often do not affect a high proportion of the pathogenic fungus unless the humidity and temperature are high; less than 50% is common (Fig. 3.13). As the spore-producing pustules of the rust are particularly invaded, this may reduce inoculum production (e.g. of uredospores) though there is really plenty of the pathogen left to cause damage to the plant. Furthermore, it seems that the mycoparasite only becomes active, or at least is only noticed, in the presence of considerable amounts of disease; this would not be an acceptable control measure if it only exploited high levels of infection. It may be possible to use mycoparasites under the high temperature and humidity conditions of some parts of the tropics or in glasshouses with high levels of introduced inoculum. *A. quisqualis* shows some promise as part of an integrated control system for mildew in glasshouses. In experimental systems at least, the high levels of introduced inoculum are not usually maintained and the potential

Fig. 3.12. (*a*) Disease severity on soybean leaves inoculated with different strains of *Bdellovibrio bacteriovorus* mixed with the pathogen *Pseudomonas glycinea*. (*b*) Percentage of the plants showing systemic blight symptoms when inoculated as (*a*). (From Scherff, 1973.)

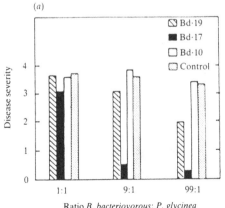
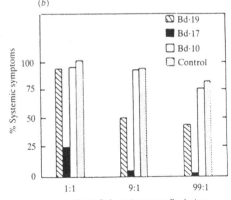

control agent dies out even under favourable conditions. It looks unlikely that there will be useful control under temperate field conditions, especially as these are situations where in general there is good disease control by varietal resistance and chemical means.

3.7   Ice nucleation bacteria

This section concerns the prevention of an abiotic disease, freezing damage, rather than the control of a disease caused by fungi or bacteria. Citrus trees, some rosaceous trees such as pears, and many crops such as potatoes and strawberries are either grown near the limit of their range so that they are not entirely hardy in the prevailing climatic conditions, or there is a premium price for early produce so that they are planted too early in the year. Such crops are likely to be damaged by frosts in the late spring and an attempt is now being made to reduce injury, by biological means, which will eventually involve the release of genetically engineered organisms into the environment. This has become a test case for such organisms and somewhat of a *cause célèbre*.

Ice forms when the water temperature drops below 0 °C provided that there are ice nucleation sites to initiate crystal formation. Such sites may be abiotic, but it also turns out that many bacteria (strains of *Pseudomonas syringae*, *P. fluorescens* and *Erwinia amylovora* the cause of fire blight, section 6.2) also act as ice nucleation sites. In the absence

Fig. 3.13. Frequency of *Darluca filum* as a % of *Puccinia cynodontis* affected in relation to leaf age (number 1 oldest) of *Cynodon dactylon* and the numbers of sori. (From Kranz, 1981.)

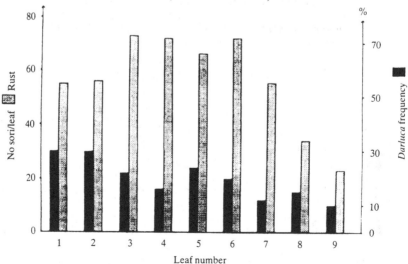

of such crystal initiation sites the water can supercool, remaining liquid down to several degrees below zero (−5 or even −12°C) so that there is no ice formation and therefore no frost damage, even during a moderate air or ground frost. If the bacteria can be altered so that they no longer act as nucleation sites, or if they are replaced by other inactive bacteria, then ice damage will be reduced. Strains of *P. fluorescens* and *P. putida* have been shown to be effective at suppressing ice damage, but we will concentrate on the more extensively investigated *P. syringae*.

Antagonists can colonize young leaves, and flowers, for one or two months after application; only a short time is needed while there is a risk of late spring frosts. It is important to establish the antagonist early before a natural colonization by the detrimental organisms. Some data on the population levels of three antagonists are shown in Fig. 3.14 where the reduction in the ice nucleation bacteria occurs with two of the

Fig. 3.14. Total bacteria (○), *Erwinia amylovora* (■), ice nucleation active bacteria (△) and ice nuclei at −5 °C (□) on leaves and flowers of Bartlett pear trees that were (*a*) untreated; or (*b*) sprayed with rifampicin resistant antagonistic bacterium strain A517 x−−−x; or (*c*) strain A511 x−−−x; or (*d*) strain A510 x−−−x. The vertical bars represent the standard errors of the mean log populations. (From Lindow, 1985.)

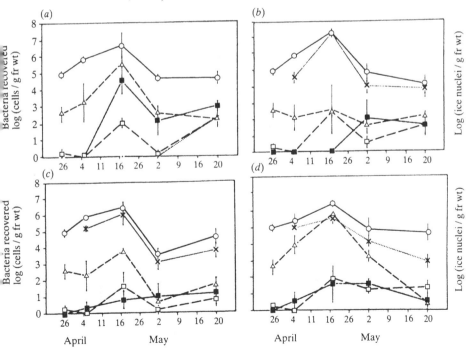

potential antagonists (*b* and *c*). These antagonists reduced the leaf populations of the pathogen *Erwinia amylovora* and the amount of frost damage to pear fruits produced on the protected trees.

Only a little over half of the antagonists to the ice nucleating strains of *P. syringae* seem to produce antibiotics in culture and it was concluded that antibiotic and siderophore production was not a prerequisite for antagonism (Fig. 3.15). Competition was probably important and this was shown by the use of 'near isogenic strains of *P. syringae*' (Lindow, 1985) which were deficient in the ability to be ice nucleation sites. Such bacteria competed with the wild type, ice nucleating *P. syringae*, and gave protection from sub-zero temperatures. So it is colonization and competitive exclusion that is important in the biological control of frost damage, rather than the production of antibiotics and siderophores. An isolation and selection procedure based on *in vitro* antagonism tests would probably not have produced useful antagonists.

Ice nucleation deficient strains (ice⁻; ice minus) strains of *P. syringae* have now been produced by genetic engineering (as opposed to selection after using non-specific chemical mutagens). The ice nucleation compound is a membrane bound protein and the genes have been

Fig. 3.15. As 3.14 but the tree was sprayed with antagonist isolate A511–6 (x – – – x) which is a mutant selected so as not to produce antibiotic in culture. Since there is still control it suggests that antibiotics are not important in the interaction. (From Lindow, 1985.)

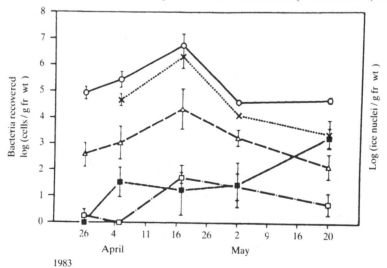

partly sequenced and cloned into *Escherichia coli*. There are now available ice⁻ strains where it is known that only this characteristic has been deleted and it is proposed to test these in the field as they have shown good responses in laboratory studies. The release of genetically engineered bacteria has been opposed in principle, even though in this case it is a deletion of a specific activity rather than the release of an organism with any extra unnatural capability. Indeed ice⁻ *P. syringae* from genetic engineering is more like the wild type than the chemically induced mutants, or even naturally occurring and selected mutants, where essentially unknown changes have been made from the wild type. The US Environmental Protection Agency and all other relevant safety and agricultural bodies have cleared the ice⁻ *P. syringae* for experimental release but the experiment was blocked by various groups against the release of such organisms or the use of genetic engineering in general. The experiment was finally done in the summer of 1987, having been first proposed in 1982. Such is the strength of feeling and differences of opinion in this matter that subsequent trials have been vandalized and damaged. though they have now (1989) been completed.

This biological control of frost damage has several interesting implications. Firstly, it is one of the few biological procedures for leaves that is protected by patents, is under commercial development and looks as though it will eventually be marketed commercially, even if only with natural antagonists or artificial mutants. This is partly because it does something which is not already done better or cheaper by chemical or other means; there is an obvious open market for such a measure on a worldwide basis. Secondly, and in the long run more importantly, it has become a test case for the use of genetically engineered organisms in agricultural plant protection. There is, rightly, much concern over the possible release of genetically engineered organisms and most countries have control agencies enforcing strict, if not always logical, codes of practice. There is less problem with the release of natural organisms back into the environment (section 2.6). It is also possible to release mutants that have been selected or even produced *in vitro*, where the genetic differences from the wild type may not be clearly defined, and will certainly not be as precise as the changes in genetically engineered microbes. Such selected and commercially produced organisms are freely released at present in the case of *Rhizobium* and *Agrobacterium*. The release of genetically engineered organisms will come in the foreseeable future, but if the case of ice nucleating *P. syringae* is anything to go by it will be a considerable trouble to test and to justify each one. The general problem of testing is beset by methodological problems and limitations: it is at present much more difficult, if not impossible, to track particular microbial strains in the environment, than it is to find a pesticide or its breakdown product(s) where the methods have been developed for many years. There is the added complication that a biocontrol agent could multiply in the environment as opposed to a chemical agent which may change, but where the concentration usually decreases with time.

3.8   Conclusions on the biocontrol of leaf diseases

There is no doubt that biological control of leaf pathogens does occur under laboratory conditions and even in the glasshouse and field occasionally. If nothing else, it is shown to be functioning by the effects of fungicides on non-target organisms. There are great possibilities for selecting or producing potentially useful organisms.

There are, however, three major problems. Firstly, survival is often poor, never mind growth and spread, and is limited especially by lack of water and nutrients in most environments. Secondly, there is the problem of relatively poor disease control, leaving a significant amount of disease even where the biocontrol 'works'. Thirdly, there is the problem, especially serious with leaf diseases, of competition with chemical control and varietal resistance. Both may break down and the former may be limited by legislation now or in the future.

The limitation on biocontrol on the leaf surface is not the biology, especially with the prospect of genetically engineered organisms, it is imposed by the environment outside or by the prevailing public opinion, and legal and financial restraints. At the moment biocontrol on leaves does not just have to work, it has to be good, very good, to stand a chance of ever being seriously considered. However, there is hope for the future when these limitations may change and new techniques of inoculum formulation may allow longer periods of survival or growth on the leaf surface.

4

Biocontrol of stem diseases

4.1   Introduction

There are some diseases that clearly can be described as stem diseases such as wound infections, timber decays and stem cankers on forest and orchard trees, and also wilts like Dutch elm disease. Crown gall classically infects stem bases, but may also cause galls on roots. We will consider all of these in this chapter, but there are some stem base diseases that are excluded and will be dealt with in Chapter 5 on roots: these include the cereal stem base complex of eyespot, sharp eyespot and *Fusarium* which are all trash- or soil-borne diseases. Similarly the seedling diseases such as damping off, which may affect stems, are considered in Chapter 7, and fire blight, which eventually causes death of stems, is described under infections of flowers (Chapter 6).

Most of the infections that we will consider are therefore of woody stems or twigs. There are rather few diseases of herbaceous annual stems, possibly because, in comparison with leaves, they are hard to penetrate and rather low in nutrients. Woody stems are a very specialized habitat, and one in which biological control has been most effective. They are generally covered by waterproof bark, rich in tannins and phenols, which successfully excludes most organisms. There are fungi and some bacteria growing on bark but they use it mainly as a growing surface and derive very few nutrients from it. The wood itself has very low numbers of micro-organisms in it when young and healthy and it is therefore relatively easy to introduce inocula and have colonization since there is very little competition. Eventually there is infection through wounds, dead branches or roots and the wood is colonized by decay organisms, especially wood-rotting basidiomycete fungi. The interactions between different fungi can be very complex

(Cooke & Rayner, 1984; Rayner & Boddy, 1986). Protection of the wood can be achieved by protecting the relatively small and well defined wound or branch stub and this again favours some biological control measures. At the initial colonization there are some free sugars and non-structural carbohydrates which are important nutrients in establishing the succession; potential antagonists such as *Trichoderma* can successfully compete for these and so control the development, or otherwise, of the succession.

So if we are looking for a site for biological control, the woody stem is an obvious choice: it is a 'clean' environment without competition, there are some nutrients initially available at defined wound sites where infection normally takes place and a considerable amount is known about existing colonization patterns and the organisms involved. Most of the successful, commercially available biological control systems are from diseases of woody stems.

4.2   Fungal wound infection of trees

There are several reports of wound floras of forest trees which seemed to be antagonistic to wood rotting hymenomycetes, but these have not been developed commercially. The control of silver leaf on some fruit trees (caused by *Chondrostereum purpureum*) by the use of dowels or pellets colonized by *Trichoderma* is commercially available, but it is a small, specialized market. There are, however, major diseases in this class that have commercially available biocontrol measures based on a number of different mechanisms and we will now consider these in more detail.

Table 4.1 *Effect of different dosages of* Fusarium lateritium *conidia on infection of pruned apricot sapwood with* Eutypa armeniacae

Protective treatment	Total stems harvested	No. infected with *E. armeniacae*	Percentage infected
Nil	49	24	49
F. lateritium:			
10^4 conidia/ml	49	23	47
10^5 conidia/ml	48	20	42
10^6 conidia/ml	50	6***	12

*** Significantly different from control ($P = 0.001$); $\chi^2 = 18.89$.
Carter, M. V. & Price, T. V. (1974). *Australian Journal of Agricultural Research*, **25**, 105–19.

4.2.1   Eutypa *and* Nectria

Much work has been done on pruning wounds to orchard trees, especially apricots, which are infected by *Eutypa armeniacae*. *Nectria galligena* infects apples through leaf scar wounds. Both these ascomycetes cause cankers on the stems and eventually death.

Eutypa can be controlled by benzimidazole fungicides, but high levels are needed and the protection does not last long. The wounds are naturally colonized by *Fusarium lateritium* which was shown to produce a non-volatile, water soluble antibiotic in culture which inhibits spore germination and growth of the *E. armeniacae*. Conidia of *F. lateritium* are applied to the pruning wound with specially adapted cutters, which deposit the spore suspension as the cut is made. The effect is dependent on the spore density and at least 10^6 conidia per ml are needed (Table 4.1) in the initial inoculum, though the antagonist subsequently sporulates on the wound surface and continues the protection (Fig. 4.1). A further refinement of this technique was possible because the *Fusarium* was ten times more tolerant than the *Eutypa* to the benzimidazole fungicides such as benomyl: integrated control was therefore possible and gave better results than either the fungicide or the biological method alone. The standard application that is now

Fig. 4.1. Sporulation of the antagonist *Fusarium lateritium* on pruned apricot sapwood after inoculation with a suspension containing 10^4 macroconidia per ml. This increases the protection against *Eutypa*. (From Carter & Price, see Table 4.1.)

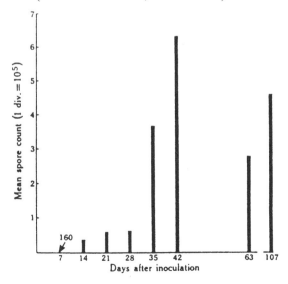

recommended in Australian orchards is 10^6 conidia of *F. lateritium* per ml plus 200 µg/ml of benomyl with the special delivery system to ensure application to the correct place as soon as the wound is made. The benomyl gives immediate protection and the *Fusarium* a longer lasting effect, in addition the fungicide can be used a+ a lower concentration than is needed if it were used alone (Fig. 4.2). This is a good example of integrated control combined with the correct delivery system achieving better control than either component alone.

For the control of *N. galligena* a variety of saprotrophs, including *F. lateritium*, were isolated from leaf scars and some were antagonistic in culture, but most did not work in the field. The best organism was *Bacillus subtilis*, which produced two antibiotics in culture and persisted on the leaf scars throughout the winter and early spring, reducing the percentage of shoots with cankers from 26.3 to 17.5%. Other workers have isolated antagonistic fungi such as *Cladosporium cladosporioides* which gave a 65% reduction in cankers. None of this is in commercial use as yet, for apples are sprayed so many times for *Venturia* (section 3.6.1) that *Nectria* is controlled anyway.

Fig. 4.2. Biological and chemical protection of apricot pruning wounds in response to treatments with benomyl and *Fusarium lateritium* at 10^6 conidia per ml. (From Carter, M. V. (1983). *Australian Journal of Experimental Agriculture and Animal Husbandry* **23**, 429–36.)

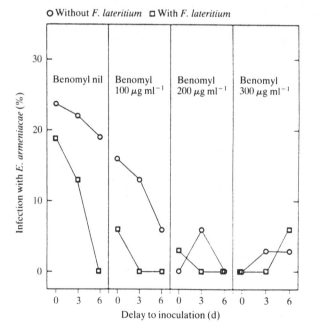

4.2.2 Butt rot of conifers

One of the first commercially available biocontrol agents was *Peniophora* (now called *Phlebia*) *gigantea* for the control of *Fomes annosus* (= *Heterobasidion annosum*) on pine trees (Rishbeth, 1963). It does not work so well on spruce trees, though other fungi have been tried for this genus, including *Trichoderma*. The butt and stem rot caused by *Fomes* is especially a disease of managed plantations and may cause an average loss of up to 10% of the annual growth of temperate conifer plantations. It is important to understand its life cycle and epidemiology to control the disease. The long range dispersal is by basidiospores, which germinate on fresh wound surfaces, usually stumps left by thinning, and it then invades the dying root system from where it can attack neighbouring healthy trees through natural root grafts, causing group dying. Once it is in a plantation in the root system it is impossible to eradicate, so control depends on keeping it out as much as possible. *Fomes* is a very poor competitor and will only invade freshly cut stumps; it is easily prevented from colonizing by a variety of antagonists, of which *Peniophora* occurs naturally and is the most effective (Table 4.2). The first control measures used were chemical: creosote was applied to the stump to protect it from invasion. This was effective, but it kept out the natural antagonists as well as *Fomes*, which could still invade along roots if it was already in the plantation. More

Table 4.2 *The extent of colonization of Scots pine stumps at 2 intervals after inoculation with various fungi as antagonists to* Fomes *which was inoculated onto all stumps*

	Mean % areas of stump section colonized after					
	10 weeks			6 months		
	Species inoculated	Pg[a]	Fa[b]	Species inoculated	Pg	Fa
None	—	28	38	—	80	7
Botrytis cinerea	5	5	55	0	0	25
Trichoderma viride	0	10	65	0	43	40
Leptographium lundbergii	95	0	5	37	47	0
Peniophora gigantea	80	80	Trace	75	75	0

[a] *Peniophora gigantea.* [b] *Fomes annosus.*
From Rishbeth, 1963.

recent chemical treatments have included urea and ammonium sulfamate which kill the stump and, by supplying nitrogen, encourage saprotrophic growth especially of *Peniophora* which colonizes the whole stump and denies the roots as a food base for *Fomes*. *Peniophora* will also replace any *Fomes* by hyphal interference and the production of short range antibiotics (Fig. 4.3). The next obvious step was to inoculate with the *Peniophora* itself and it turned out that very little inoculum was needed on pine stumps (Table 4.3): there can be 15 or even 50 times more *Fomes* than *Peniophora* inoculum and you still get control. The dry pellet of *Peniophora* spores has a shelf-life of two months at 22 °C, or a wet spore suspension will last four months, as long as the temperature is less than 20 °C. The material is diluted in water before application. Provided that the stump is freshly cut *Peniophora* will colonize well (Table 4.4) and will persist: it is easy to introduce an antagonist into a virtually sterile environment. There are also some indications that *Peniophora* may be used to colonize spruce and Douglas fir stumps when combined with nitrogen additions (ammonium sulphate and ammonium sulfamate). This is again a relatively small specialized market for a biocontrol agent but one where it was possible to target the antagonist carefully, to take advantage of a virtually sterile infection court and to use the agent as a protectant as well as an eradicant.

Table 4.3 *Colonization of pine stumps by different proportions of mixed inocula of* Fomes *(Fa) and* Peniophora *(Pa)*

	Mean % areas of stump section colonized after 4 months					
Approximate dosage ratio Fa:Pg	(A) Fa dosage 1×10^5			(B) Fa dosage 4×10^3		
	Pg dosage	Pg	Fa*	Pg dosage	Pg	Fa*
Fa only	0	0	23 (7)	0	0	22 (7)
50:1	2×10^3	97	<1 (1)	80	76	8 (4)
15:1	6.7×10^3	99	0	2.7×10^2	83	1 (3)
5:1	2×10^4	97	0	8×10^2	84	2 (1)
1.5:1	6.7×10^4	96	0	2.7×10^3	73	0
0.5:1	2×10^5	100	0	8×10^3	96	0
0.15:1	6.7×10^5	100	0	2.7×10^4	95	0

* The number of stumps out of 10 containing *F. annosus* is given in brackets.
From Rishbeth, 1963.

Table 4.4 *Natural colonization of Corsican pine stumps by* Fomes *after inoculation with different* Peniophora *isolates*

Peniophora isolate	Area of stump section (%) colonized after 8 months by	
	F. annosus[a]	*P. gigantea*[a]
None	20	55[b]
A	1	87
B	0	81
C	1	70
D	0	66

[a] Mean of 12 replicates.
[b] Natural infections.
From Rishbeth, J. (1975). In *Biology and control of soilborne plant pathogens*, ed. G. W. Bruehl, pp. 158–62. St Paul, Minnesota: American Phytopathological Society.

Fig. 4.3. Hyphal interactions between *Fomes* (growing left to right) and *Peniophora* (growing top to bottom). The *Fomes* has granular cytoplasm and has taken up the stain, indicating a damaged plasmalemma.

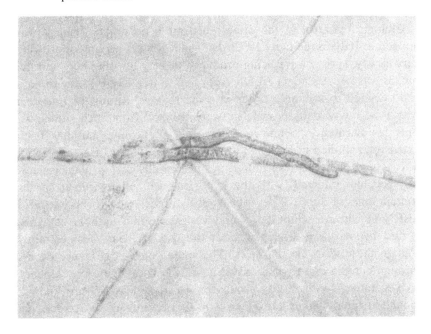

4.2.3   Chestnut blight

The case of chestnut blight, caused by the ascomycete *Endothia parasitica*, is different again, for here we are dealing with an introduced pathogen which, though requiring a wound, normally uses natural damage rather than man-made cuts. The mycelium grows under the bark and forms a canker on the branches and the stem which eventually girdles the stem and kills the tree. It can grow saprotrophically on chestnut and some other species, so there is a constant pool of inoculum even when the primary host has been killed. The cankers bear pycnidia which produce large quantities of asexual spores, and eventually perithecia are also produced. These spores are spread by birds, insects and rain splash.

Endothia seems to have originated in Asia and the chestnuts there have some resistance to it, but American and European chestnuts have very little resistance, though less virulent isolates may be temporarily walled off by zone lines in very vigorous trees. There are no effective fungicides, certainly none that are economically feasible. The fungus was introduced into North America from Asia in 1904 and is probably the most devastating plant disease known (Cook & Baker, 1983): it killed all mature chestnuts on the eastern seaboard within a few years and now only stool shoots regrow, but are themselves infected as soon as they get more than 10 cm in diameter. It was introduced into Italy in 1938 and later into France.

Biological control of the disease became a possibility when it was noticed in Italy in the early 1950s that cankers were healing or growing very slowly, spores were being produced in very low numbers and the fungus was not spreading rapidly from tree to tree. By 1978 the disease had declined to tolerable levels in some regions. Strains of *Endothia* with low, but variable, virulence were isolated from such situations. This hypovirulence is known from other pathogens, but has been extensively studied in *Endothia* (Anagnostakis, 1982; van Alfen, 1982). The hypovirulence is transmitted cytoplasmically and virulent isolates will become 'infected' with the hypovirulence. Hypovirulent strains contain one or more types of double stranded ribose nucleic acid (dsRNA) within membrane bound vesicles. Some authors have reported hypovirulent strains without the dsRNA, but most certainly contain these virus-like particles. The spread of hypovirulence occurs through conidia and hyphal anastomosis; the latter therefore depends on vegetative compatibility groups of which at least 77 are known, though fortunately only a few (less than 10) occur in any one region of a country. There are at least five loci defining the compatibility groups and the ease of transmission increases with increasing numbers of loci bearing different alleles. In Europe hypovirulence has been used commercially for the control of *Endothia*. Suitable hypovirulent strains, to match the local vegetative compatibility types, are introduced into the region. Usually several different strains are used at once and possibly only ten inoculations per hectare are needed. The hypovirulent strains spread from these foci and can eliminate active cankers within ten years. Large areas are now being treated in this way at a cost of millions of dollars. In North America the situation was slightly different and there were initial problems with biological control. There are strains with varying degrees of virulence which, especially in mixture, can limit the size of cankers (Figs. 4.4 and 4.5) when inoculated in the field, but the problem was with getting the correct compatibility groups. There are now American isolates from naturally occurring hypovirulent strains. Since it can grow as a saprotroph, and occasionally hypovirulent strains revert to virulence (e.g. from ascospores), it is unlikely that the

Fig. 4.4. Pathogenicity of single and mixed cultures of four hypovirulent strains of *Endothia parasitica* on excised stems of American chestnut. Bars are standard errors. (From Jaynes, R. A. & Elliston, J. E. (1980). *Phytopathology* **70**, 453–6.)

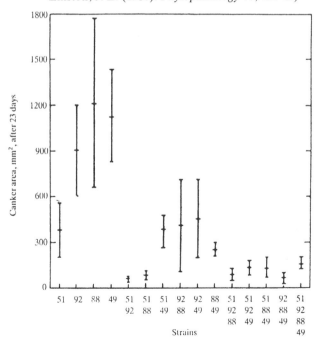

be totally eliminated, but biological control by the use of hypovirulence looks as though it will be successful.

4.2.4   Dutch elm disease

This is another example of the problems that can arise with diseases introduced to host populations outside their normal range. Dutch elm disease (which is no fault of the Dutch, they were simply the first people to take the trouble to investigate it!) is caused by an ascomycete *Ceratocystis ulmi*, which is spread by particular species of bark beetle (*Scolytus scolytus* and *S. multistriatus* in the UK). The beetles inoculate their brood galleries with the fungus and the emerging adults are infected: they then feed on young bark of the twigs, and so infect other trees, before inoculating their own brood galleries. The disease has been around for many years, and there are known to be strains with varying virulence. Particularly virulent strains appeared in North America and Rumania and from there have spread or been imported into Britain and the rest of Europe where they have virtually wiped out the mature elms; though suckers and stool shoots are regrowing they will probably also become infected. Again there are no effective, economically feasible

Fig. 4.5. Cankered area and % of American chestnut stems surviving two growing seasons after infection with normal (control) and hypovirulent strains of *Endothia parasitica*. '8 around' is 8 individual strains placed around a canker. (From Jaynes & Elliston, see Fig. 4.4.)

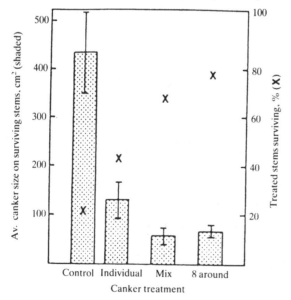

fungicides, though some of the systemic ones have been tried with varying success. A whole range of elm bark saprophytes have been isolated (e.g. *Pseudomonas, Trichoderma, Gliocladium*) and show some promise, but they have not been tested in the field. *Phomopsis oblongata* colonizes the bark soon after infection by *C. ulmi* and is detrimental to the beetle larvae, so preventing the spread as well as inhibiting the pathogen. The larvae fail to grow and pupate (Fig. 4.6). There are also reports of dsRNA infections of *Ceratocystis* that are linked to cytoplasmically transmitted hypovirulence as in the control of chestnut blight. So Dutch elm disease continues to be damaging and is still being intensively investigated but effective field control by biological means remains a possibility rather than a reality.

Fig. 4.6. (*a*) Breeding galleries in elm bark of *Scolytus multistriatus* in normal condition. (*b*) Sparse and abnormal breeding galleries after the elm was colonized by *Phomopsis* as a control agent for Dutch elm disease carried by the beetle. (From Webber, J. (1981). Reprinted by permission from *Nature* **292**, 449. Copyright © 1981 Macmillan Magazines Ltd.)

(a)   (b)

4.2.5   Silverleaf disease

This is a disease which affects fruit trees, especially pears and plums. It is caused by *Chondrostereum purpureum* growing in the stem and producing a toxin which causes air spaces to form in the palisade mesophyll when translocated to the leaves, giving a silver appearance. The stems can be inoculated with *Trichoderma*, grown on wooden dowels or prepared as pellets which are inserted in holes bored in the stem. Infected trees show increased rates of recovery from the disease compared to uninoculated controls (this is recovery from existing infection, not protection from disease). Alternatively the *Trichoderma* can be used as a protectant to prevent establishment of the *C. purpureum* on pruning wounds.

4.3   Diseases of herbaceous stems

Successful work on the biological control of herbaceous stem diseases has been done on carnation stem rot, caused by *Fusarium roseum* 'Avenaceum'. A non-pathogenic *F. roseum* 'Gibbosum' was used to pre-inoculate wounds during propagation and it produced a germination inhibitor and also reduced the time until the stems developed resistance to the rot. This was not antibiosis or competition, and the fact that autoclaved culture filtrates and inorganic ions, such as mercury and copper, had a similar effect (Fig. 4.7) led to the suggestion that the biocontrol agent was working by stimulating the host's phytoalexin defence mechanisms. This is, therefore, an example of the type of cross-protection known as induced resistance (see section 3.6.2) in which one fungus, by stimulating the defence system gives protection against the pathogen. See also Fig. 7.4 for carnation stem diseases.

4.4   Crown gall

Crown gall is one of the main examples of a bacterial stem disease. There are others such as blackleg of potatoes, caused by *Erwinia carotovora* subsp. *atroseptica*, for which there are reports of some control by inoculation with *Pseudomonas* but not nearly so much work has been done on this. Crown gall is mainly a disease of nursery planting stock and propagation material where the wounds on cuttings and grafts provide an infection court. It is particularly important on peach, plum, almond and other fruit trees, and also on vines and to a lesser extent on herbaceous stems like chrysanthemum: altogether it is known to infect 93 families of plants and the world losses are estimated at more than US $138 million (Moore & Cooksey, 1981; Thomson, 1987). Galls typically form on the crown of the plant (where the stem enters the soil) and on the roots and above-ground stems: they have even been reported from the leaves in rare cases. The causal organism is *Agrobacterium tumefaciens* of which there are various biotypes with different host specificity and different degrees of virulence. It can be found in soil, though it does not survive there for more than a year or so if introduced, and the main source of infection seems to be galled plants. Spread through irrigation water has been recorded, but infection usually occurs from the soil and rhizosphere and from knives, cutters, etc. during propagation where infected stock plants have been used.

Virulence of *A. tumefaciens* is basically determined by the possession of a plasmid (an extrachromosomal portion of DNA which can replicate independently and may be transferred from one organism to another) called the Ti-plasmid. Strains of *A. tumefaciens* can gain and lose virulence by acquisition or loss of the plasmid. Products from the susceptible plant enter the bacterium and activate virulence genes on the plasmid. Part of the Ti-plasmid is then transferred in an unknown

Fig. 4.7. Effect of culture filtrate of *Fusarium roseum* '*Gibbosum*' and various chemicals on the subsequent infection by *F. roseum* '*Avenaceum*'. (From Baker, R. *et al.* (1978). *Phytopathology* **68**, 1495–501.)

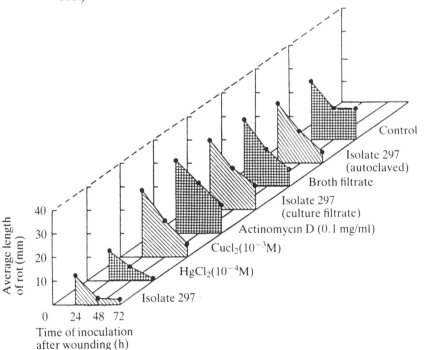

way from the bacterium to the eukaryotic host cell and there integrates with the host DNA and is expressed (Fig. 4.8). The plasmid carries several different genes or gene sequences which result in specificity, in the production of auxins and opines by the host, and may also be responsible for the sensitivity of the bacterium to some antibiotics.

There are specific binding sites on the host walls to which the bacterium attaches by lipopolysaccharides, which are probably coded on the plasmid, though there is evidence that the main bacterial chromo-

Fig. 4.8. Ti-plasmids and pathogenesis: a schematic representation. Plant products enter *Agrobacterium* cells and induce virulence genes (VIR) and on the Ti-plasmid (pTi). Transfer DNA (T-DNA) forms circles joined at T-DNA extremities by one terminal repeat sequence (TS). The T-DNA is transported to the plant cells and integrated in the plant chromosome. Oncogenic genes (ONC) and opine synthesis genes (OPS) are expressed in the plant cells. Abnormal levels of plant growth regulators produce galls or hairy roots, and then tissues release opines which are used by the *Agrobacterium* either to grow or as conjugation inducers which promote plasid multiplication and dispersal. (From Clare, B. G. *et al.* (1987). In *Genetics and plant pathogenesis*, ed. P. R. Day & G. J. Jellis, pp. 79–90. Oxford: Blackwell Scientific Publications.)

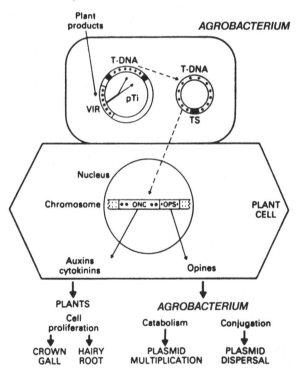

some is also involved. Killed or avirulent bacteria will bind to these sites and exclude the virulent strains and the amount of attachment is also determined in part by the general chemistry of the host wall, especially the amount of methylation on the polygalacturonic acid units of the pectin polymer.

The genes responsible for host cell proliferation (oncogenic genes) cause the synthesis of both cytokinins and indole acetic acid within the plant cell (Fig. 4.8) and the galls, or sometimes extra roots, are formed.

Opines are substituted amino acids known to be produced only by plant cells infected by the Ti-plasmid and they can only be metabolized by *Agrobacterium*, so the host produces a unique food source for the pathogen, which may then produce more cells and more Ti-plasmids. A considerable number of different opines are known and different strains of *Agrobacterium*, with different host ranges, carry Ti-plasmids which code for the production of different opines.

Some bacteria produce antibiotics that are effective against closely related bacteria, e.g. within the same genus. These special antibiotics, called bacteriocins, are usually proteins and are effective in very low concentrations. *Agrobacterium radiobacter* strain K84 is non-pathogenic and produces a bacteriocin called agrocin 84 (which is an adenine derivative, not the usual protein). It is active especially against those strains of *A. tumefaciens* that attack stone fruits and which produce the opine agrocinopine A. The uptake system for this opine is utilized by the bacteriocin, so the pathogen has a specific system that takes up the agrocin 84 which kills it.

This is a rather complex story which has been much investigated because the Ti-plasmid is one of the few ways known of getting prokaryotic DNA into eukaryotic higher plants. It has therefore been used in genetic engineering to introduce novel genes into plant cells, but what has all this to do with biological control? Crown gall, especially on stone fruits, is effectively controlled by commercially available inocula of *A. radiobacter* strain 84, which produces agrocin 84. It does not work against all crown gall infections because strains other than those producing agrocinopine A and related opines are not sensitive to the bacteriocin. There are, however, other modes of action, including the saturation of adsorption sites with non-pathogenic strains. New strains are also under investigation which are genetically manipulated so that they cannot transfer the plasmid, cannot become resistant to agrocin 84 or have deletions of some vital parts of the Ti-plasmid. The genes for production of agrocin have also been transferred to other bacteria such as *Rhizobium*. Strains of *Agrobacterium* producing other bacteriocins are also known. These strategies should broaden the host range of the biocontrol agent and increase its already considerable use worldwide.

In simple terms, all the grower has to do is to dip his plants or cuttings in a suspension of the biocontrol agent during propagation or transplanting. The knowledge needed to produce a biocontrol agent and the mode of action of that agent, can be as complicated and esoteric as you like, but in use it must be cheap, simple and effective.

4.5   Conclusions

We should pause a moment to stress again the importance to the general development of biological control techniques of the diseases described above. They are not in major world food crops like cereals, potatoes or legumes but in rather specialized situations or in tree crops. They do serve, however, as examples of what can be done by biological control and they are the hope that inspires others who are working on apparently more difficult diseases or crop situations. So the question that must be asked is: what is special about woody perennials or stem diseases that makes biological control work here and not elsewhere? If there is something special, then the hope for general use is misplaced, but there may be other reasons why successful biological control just happens to occur under these conditions.

At first sight there is nothing special about perennials except that, as pointed out in the introduction, the large stems are microbiologically fairly simple and are good sites for inoculation of antagonists. When this is not true, as in the case of *Agrobacterium* or carnations then the horticultural procedures allow for especially effective inoculation, by dipping the whole plant. It is not possible to do this on field grown crops like wheat. So the first answer is a cautious yes, the diseases discussed above do seem favourable for biological control. However before we get too depressed let us look again at the major world crops.

Just because they are major crops, and therefore major markets, the diseases of cereals etc. have been extensively studied by plant breeders and by the agrochemicals industry. For better or worse in the late 1940s it was thought best to control plant diseases (and pests) by the use of host resistance, and by the use of biocides which industry had learned how to produce on a commercial scale. Enormous resources have been put into this (see section 2.2) and very effective control measures have been found. It is speculation, which will never be answered, whether or not equivalent resources devoted to biological control would have produced such useful results, but the very success of plant breeding and chemical methods has until recently discouraged biological control investigations on a reasonable scale. So maybe major crops have not been successfully worked on for biological control because, in general, there were other effective methods available.

Conversely, if we look again at the biocontrol of diseases described above they are on crops that have not been worked on by plant breeders for resistance or by the chemical companies for pesticides. Plant breeders must work on very long time-scales for progeny testing perennials, until some of the recent tissue culture techniques were introduced anyway. Breeding for disease resistance in trees is not common. In addition they are marginal markets; at best they use pesticides whose development costs have been justified by major crops, but which have subsequently been found useful in other situations, perhaps with some relatively minor reformulation. In these situations biological control methods, with introduced antagonists, have been viable alternatives and have been developed beyond the experimental, laboratory stages which characterized the biological control systems for leaf diseases, for example (Chapter 3).

So to answer the initial question, there are specially favourable circumstances that have encouraged successful biological control of the crops and diseases of stems discussed in this chapter. The reasons why biological control methods have not been developed for many major crops are economic and historical. It is possible to use these good examples on stems to encourage work on biological control of diseases of other plant organs and of major crop diseases.

5

Biocontrol of diseases of roots

5.1　Introduction

There is much information on the microbiology of soils (Nedwell & Gray, 1987), especially the soil near roots which will mostly concern us. Soils are very variable on many different scales. There are the differences that occur between soil types, usually based on the parent material, the climate and the vegetation, which control the amount of clay, organic matter and so on. Secondly, there are differences within soil in relation to depth, which reflect the addition of organic matter to the surface and are the result of leaching down the profile: the soil may be divided into a number of layers (called horizons) which have very different physical and chemical characteristics. Thirdly, there are differences on a very small scale, the microhabitat, which reflect changes in nutrient status, substrate availability, aeration, etc. on different parts of a soil crumb or a sand grain: we may here be talking about distances of a few tens of micrometres making a significant difference in oxygen levels because of the very low solubility of this gas in water (Campbell, 1983; Bruehl, 1987). There is, therefore, great variation in microbial numbers and activity between and within soils. There may be several million bacteria and hundreds of thousands of fungi which can be cultured from a gram of soil, but many of these will be inactive in the soil because of the environmental limitations which most commonly are temperature, water availability, aeration and available substrates for metabolism and growth. Almost all soils are carbon limited for heterotrophic microbes; even though they contain organic matter this is often not available because of spacial limitations (the organisms cannot get at it) or they do not possess the correct enzyme systems to degrade it.

An exception to this nutrient limitation is the region around plant roots, the rhizosphere (Curl & Truelove, 1985; Campbell, 1985), where simple sugars and amino acids, and many other compounds, are exuded by the plant and are available to the micro-organisms. There are also dead cells from the cortex and the root cap which, together with the mucilage on the root, form a less readily available source of nutrients. In the rhizosphere the normal carbon limitation is therefore removed and there is an increase in the numbers and activity of many sorts of micro-organisms, including plant pathogens. Growth may be stimulated or directed towards the root, and spores and sclerotia may germinate in response to exudates in general or specific components of them. There are changes from the normal soil apart from the availability of nutrients: the extra activity of heterotrophs may lead to low oxygen levels and higher concentrations of carbon dioxide and secondary metabolites. There may be sufficient substrate for the production of antibiotics; though these have not been shown to occur in normal soil, they can be detected when available carbon is added and they could be formed in the rhizosphere. It is in the rhizosphere that many of the interactions between root pathogens and potential control agents will occur, so it is important to understand the behaviour of microbes in this region if we are to manipulate them to reduce plant disease.

Pathogens in the soil (Bruehl, 1987) may be actively growing on and attacking roots, but the majority will be dormant, or surviving in some other stage of their life cycle, because of the limitations outlined above. Firstly, there may be dormant propagules such as sexual or asexual spores, long-lived chlamydospores resistant to unfavourable environments, or sclerotia. These are often from ruderal species (section 1.2) which will germinate, grow and infect as the opportunity arises, before they return to dormant spores in the soil. They are the unspecialized pathogens (section 1.2) *sensu* Garrett (1970). Secondly, the pathogen may survive in infected host material or trash left from the previous crop: in this case they are combative or competitive organisms defending a captured resource (the fragment of plant root or whatever) or stress tolerant organisms that are the only ones capable of surviving in, for example, a nutrient poor, rather dry piece of straw in the upper layers of the soil. Finally, there are pathogens, like *Rhizoctonia*, which live saprotrophically on general soil organic matter between their attacks on live hosts. They are often unspecialized pathogens with a very high competitive saprophytic ability so that they can survive, and indeed flourish, in competition with other organisms. When considering the biological control of a root pathogen it is essential to know this sort of information so the control can be targeted to a vulnerable stage in the life cycle.

Biological control may thus aim at a reduction in the existing inoculum of the pathogen by resource competition (starving the pathogen) or by parasitizing it as in the case of the destruction of sclerotia by *Sporidesmium* (section 5.6.1), of hyphae and sclerotia by *Trichoderma* (sections 1.3.4 and 5.6.1) or the perforation of spores or hyphae by amoebae (sections 1.3.4 and 5.7). The control agent may also restrict the germination or growth of the pathogen in the soil or on or in the root. This could be by the production of siderophores, antibiotics or other toxins (sections 1.3.3 and 5.6.3). Finally the control agent may stimulate the host defence system (induced resistance, sections 1.5, 3.6.2 and 5.5).

The biological control of diseases of plant roots has been studied since the earliest experiments in the 1920s (Chapter 2) and has been extensively reviewed in a large number of books and papers (Chet, 1987; Cook & Baker, 1983; Lynch, 1987; Parker *et al.*, 1985). The subject continues to produce an enormous number of papers in the literature, for the amount of interest has increased markedly in recent years and there is now a lot of time and effort, and therefore a lot of money, devoted to the study by both government agencies and by commercial firms. Despite this, there are at present (1988) no commercially available inocula for the control of any of the important soil-borne diseases of major field crops, though at least one is expected to be marketed within the year. Why is there all this interest in biocontrol of root diseases without much evidence of efficacy under commercial field conditions?

The plant breeders have been extremely successful in introducing or enhancing the resistance to many leaf and stem diseases in a number of crops. However, there is less work on resistance to root diseases, except perhaps for the wilts. Similarly the agrochemical companies have developed a vast range of fungicides for use on the above-ground parts of the plant, but relatively few for the roots. Even those that do exist for the treatment or prevention of root diseases are often only seed coatings or dips for transplants. There are soil drenches and fumigants but these are usually indiscriminate and powerful biocides, rather than the treatment for specific diseases. There is a continuing and extensive search for downward translocated systemic fungicides, so that the roots can be treated by spraying the leaves, but to date there are few commercially available which are effective against serious root diseases.

There are several reasons for this concentration, in the past, on the diseases of leaves and stems. Firstly, you can walk through a field or trials ground and see the symptoms, so assessment and testing is relatively easy. Root diseases may not be immediately obvious, or if they are recognized then assessment usually involves digging up the plant, so destroying the crop you are trying to protect: the result is that the trials become much more expensive as multiple plots or fields are needed to allow for destructive sampling. Secondly, the leaf and stem diseases were perceived, often correctly, as the major limitations on productivity, but many of these can now be controlled by fungicides or by the breeding of resistant cultivars. Root diseases then become the limiting factor, or are recognized as being important when the plant no longer dies from other causes. Thus Cook was able to say in 1986 that 'of all the constraints to plant health, none are more critical or overlooked more frequently than are the biotic constraints on root health'. Finally, there are many fungicides, and biocontrol agents, which are effective against the pathogens *in vitro*, but get adsorbed by soil colloids like clays and do not work in field use.

So the attention of plant pathologists has now turned more seriously to the control of these diseases. There are: (1) Major problems of many crops such as wilts caused by *Fusarium* spp., *Verticillium* spp. and some bacteria (*Pseudomonas solanacearum, Corynebacterium michiganensis*). (2) General root rots, sometimes associated with poor cultivation or wet soils, such as those caused by *Pythium, Phytophthora, Fusarium* spp. (3) Specialized rots of particular crops like take-all of cereals (*Gaeumannomyces graminis*). (4) 'Minor pathogens' of many crops: these are a rather ill-defined group of organisms, often not identified, which reduce growth and yield though they may not produce obvious symptoms. They are therefore only recognized when the soil is fumigated or drenched with fungicide to treat some other known pathogen, and the crop then grows better even when the main disease is not serious. It may be that many of the plant growth promoting rhizobacteria (section 5.8) operate by controlling these minor pathogens. There are also deleterious bacteria around roots which reduce growth by a variety of methods (section 5.8) and these may also be part of the minor pathogens.

These diseases have traditionally been controlled by cultural practices (Palti, 1981) such as crop rotation, tillage practices or addition of organic manures (section 5.3) which may operate via biological means, though this is not always proven. Thus tillage may break up the crop residues and expose the pathogens they contain to antagonists, as well as leading to more rapid breakdown of the pathogens' food base. Crop rotations prolong the time during which the pathogen must survive without its host, and reduce inoculum potential by the activity of soil microbes on the resting structures.

With so few commercially available biocontrol systems for root diseases in the field, there must be serious problems and we will meet examples of these in the following discussions. There is no doubt that micro-organisms in soil, or added to soil, can reduce the effects of pathogens on plants: many workers have shown this with a variety of diseases. The overwhelming problem is to get repeatable results (especially in the field, rather than in the laboratory or glasshouse) which are consistent from year to year and over different climatic and soil types. This variability has many causes including the sensitivity of many potential control agents to these environmental factors, especially when control depends on the growth and spread of the antagonist, as it usually does. Clays in the soil can adsorb the organism or its metabolic products and unfavourable weather can kill it. There are also the usual problems (section 2.6) with stability of cultures and the industrial-scale production of viable, effective inocula. In soil there are many other organisms that may antagonize or kill the antagonist, in contrast to some of the successful cases of biocontrol which operate in the almost sterile environments of nursery composts (section 7.1) or newly cut tree stems (section 4.1). The soil, though very variable, is also remarkably stable: the organisms present are often assumed to be a well adapted community in which there may be changes in individual species or populations but which overall remain quite constant. It is, therefore, difficult to introduce a 'foreign' organism into an environment where it does not already exist: conversely if an organism is adapted to the environment and able to exist there, then it or something very similar will already be there. A major perturbation may be needed to make such drastic changes in the environment that the newly introduced organism can find a place in the new community. A change in the cultivation, tillage or irrigation regime may be sufficient to establish and encourage the antagonist or more drastic measures such as partial sterilization by fumigation or the use of aerated steam may be necessary. These factors of survival and growth can be summed up in the ability of a potential control agent to colonize the soil or the roots of the plant and to survive for a sufficient time to protect the crop during the stages in its growth when it is susceptible to the disease. The survival or residence time is critical, but varies with the disease. For a horticultural crop, of lettuce for example, it may only need to be a few weeks, but for field crops like wheat it may need to be almost a year. Ideally an organism would be so well adapted to the environment, the host plant and the pathogen (a *K*-strategist) that it would essentially survive and flourish for a very long time; one application gives control for ever. However, from the point of view of safety clearance with environmental protection agencies it may be much easier to use an antagonist that will stop being active after a reasonably short time (an *r*-strategist): there could be serious concern about introducing an organism that appeared to survive forever in its new environment. Also for the development of a commercial biocontrol agent it is desirable to apply, and therefore to sell, the product at least once every crop, so that research can be justified by continued sales and production. This is not to say that a 'self-destruct' organism is envisaged, just that it may not be important if the organism does not last too long.

There are now several studies on the colonization and persistence of potential biocontrol agents against soil-borne pathogens. Usually mutant strains, with high levels of resistance to one or more antibiotics, are used to allow their isolation on media selective for the particular strain. The bacteria are applied to soil as a drench or seed coating and the first problem is to show that they move onto the developing root system. Thus if 10^8 cfu were put on each seed there can be at least 10^7 per g of root (Fig. 5.1, for the north western USA), so the *Pseudomonas* in this case has moved from the seed to the root and has grown enough to give high numbers on the expanding root system, which of course weighs much more than the seed, and more importantly has a very much greater surface area that has to be covered to deny the pathogen entry points from the soil. This particular antagonist also grew better in the presence of the pathogen *Gaeumannomyces graminis*. There were 3 to 15 times more antagonists on infected plants (Fig. 5.1, after day 172) though the general population of bacteria was little affected by the presence of the disease. Other work in Britain has shown a remarkably similar pattern of colonization by other strains of *Pseudomonas* which survived winter freezing but were reduced by spring drought. Antagonists vary in their ability to colonize and some may grow very slowly, and never cover the root entirely; in particular the root tip grows faster than most micro-organisms can grow or swim. Furthermore, the inoculation is usually at the top of the root system and the tip is some way down in the soil. There is also evidence that colonization by some strains may be patchy, with good cover on some roots and others with hardly any of the inoculated organism. Colonization is also affected by the environmental factors which allow growth, for example temperature, and especially by water. Water in the soil may allow a bacterium to swim or there may be bulk water movement to carry the bacteria down the roots. Colonization is usually best in wet soils above field capacity, though not wet enough to have serious problems with aeration.

5.2   Suppressive soils

Suppressive soils have been known about for nearly 100 years (Schneider, 1982). They are soils where disease development on or in the susceptible host is suppressed, even though the pathogen is present in the soil or is introduced. The mere absence of disease symptoms is not sufficient evidence for suppression. Within this broad category there have been many subdivisions, and there are undoubtedly several different processes that result in the same observation of disease suppression. Firstly, there is the suppression of disease in some soils which seems to be an intrinsic property related to the chemistry,

Fig. 5.1. Populations of *Pseudomonas* associated with roots and the remaining seed of winter wheat which had been seed treated with the bacterium at 10^8 cfu/seed. Values are given for plants with or without the pathogen (*G. graminis* var. *tritici* = Ggt). Capital letters denote months of the year. (From Weller, D. (1983). *Phytopathology* **73**, 1548–53.)

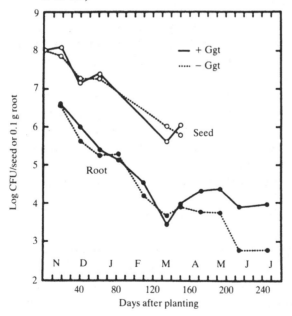

mineralogy or soil condition, such as waterlogging. Thus *Fusarium* wilt of cotton is more severe on light sandy soils than on heavier clays which are 'suppressive'.

More usually, however, the host plant and the pathogen have to be present, and the suppressiveness develops as the time under cultivation, especially monoculture, increases. This is known for potato scab (*Streptomyces scabies*), take-all decline of cereals (*Gaeumannomyces graminis*) and for *Phymatotrichum* in alfalfa and cotton. This specific suppression occurs when one particular disease is involved and it may reflect a change in the pathogen population in the soil, to avirulence for example, it may be a change in the soil properties or it may be the development of particular antagonists to the disease. A break in the monoculture, so that a non-host plant is used as part of a rotation, can destroy this effect as the pathogen is prevented from actively growing: suppressiveness of this sort needs the live, active pathogen and one of its host plants. Disease control that develops in this way is usually assumed to be microbiological in origin, and is destroyed by heating or by other means of soil 'sterilization'. Specific suppression usually involves a small number of microbial antagonists to that disease, but the change may also be more general, reflecting for example the general increase in microbial activity on the addition of readily available organic matter, and giving suppressiveness to several different diseases.

If suppressiveness is caused by micro-organisms then it follows that it can be transferred to a conducive soil (one that allows disease development) because the micro-organisms are transferred as well. The amount of soil needed in the transfer is variously reported but may be as little as 1% (w/w) in potato scab, especially when alfalfa meal is also added to encourage the growth of the relatively small inoculum. Soil transfers are also more effective if the addition is made to sterile or steamed soil so that the micro-organisms are able to colonize rapidly without competing with a resident population. For example a *Fusarium* suppressive soil was effective at 1% by weight when added to a steamed soil (Fig. 5.2*a*) but without the pretreatment over 30% suppressive soil was required to make a useful reduction in disease levels (Fig. 5.2*b*). Such a quantity would obviously be impossible on a commercial scale.

The logical step is not to bother with the transfer of the soil, which is bulky and expensive to transport, spread and mix, and which may also transfer other pathogens. It can be equally effective to transfer just the important micro-organism(s) isolated from the suppressive soil (if they can be isolated). The suppressiveness can often be induced by the

Fig. 5.2. (*a*) Incidence of *Fusarium* wilt when a suppressive soil or a conducive soil was added (1:99, w/w) to steamed glasshouse soil infested with *F. oxysporum* f.sp. *dianthi*. No new soil was added to the control. (*b*) As (*a*) but adding different amounts of the suppressive soil to glasshouse soil which had not been steamed; such soil requires higher amounts of the suppressive soil addition to control the pathogen. (From Scher, F. M. & Baker, R. (1980). *Phytopathology* **70**, 412–17.)

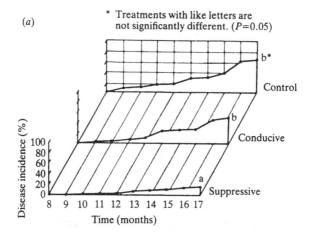

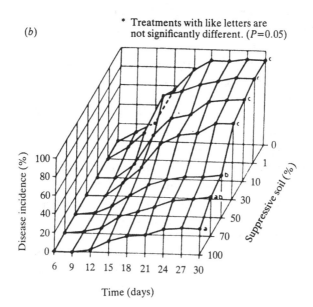

addition of micro-organisms or even in some cases by their metabolic products. Note, however, that not all potentially useful antagonists come from suppressive soils.

We can illustrate these points more clearly by taking two of the best known examples of biological control of disease, which have been grouped under the general heading of suppressive soils.

5.2.1 Fusarium *suppressive soils*

There is a carefully documented case of French alluvial soils in the Chateaurenard district of the Rhône valley where *Fusarium oxysporum* forma specialis *melonis* is present, susceptible varieties of melon are grown and the climate is suitable, but no wilt disease occurs. These soils are also suppressive for f. sp. *lycopersici, raphani, dianthi* and *cucumerinum*, but not to other species or genera such as *F. solani, Rhizoctonia solani, Pythium, Phytophthora* or *Sclerotinia* (Alabouvette *et al.*, 1979). These soils have all the characteristics of biologically suppressive soils: they lose their effect if the microbiological activity is reduced or eliminated by steaming and the ability to suppress disease can be transferred to conducive soils, though a 10% (v/v) addition is needed (Fig. 5.3) and again this would not be commercially viable. The suppressiveness does, however, appear to be an intrinsic quality of the soils for, as far as is known, it was present before the growth of any agricultural crops which are susceptible to the pathogens. Detailed analysis of the microbiology of these soils has been made and attention has centred on the fungi present. A mixture of fungi added back to the steamed soil re-introduces the suppressiveness, but individually the only fungi that have a marked effect are non-pathogenic strains of *F. oxysporum* and *F. solani* (Fig. 5.4).

The main way in which the pathogenic *Fusarium* is affected is that the growth rate is much reduced and the dormant chlamydospores do not germinate in the presence of the host root exudates as they would normally. No antibiotics have been detected and siderophores are not thought to be involved. As the effective fungi are closely related to the pathogen it was possible that the host defence mechanisms were stimulated, but there is no evidence of this. The main clue is that the suppressiveness is lost when readily available organic matter (e.g. glucose) is added to the soil: fungistasis induced by nutrient limitation is thought to be the main mechanism, with the competing *Fusarium* spp. having nearly the same niche as the pathogen.

With such a carefully investigated phenomenon, where the mode of action is apparently clear and the organisms are identified, it should now be possible to use the system commercially outside the Chateaurenard region. This will take time to develop as there are certain to be some problems with other soil types or the formulation of inoculum so as not to add nutrients which will make the competition less effective. However, this system of biological control, developed from the observations on suppressive soils, is one of the more hopeful for the future.

Other forms of *Fusarium* suppressiveness are also known. In California *F. oxysporum* f. sp. *pisi* was introduced on seed but the pea wilt only developed in two small areas. The soils had never been exposed to the pathogen and yet there was suppressiveness present. Furthermore, the suppressiveness is lost by heating or otherwise 'sterilizing' the soil and is transferable, so presumably it involves micro-organisms.

Fig. 5.3. Transmission of soil suppressiveness to a light peat by addition of 10% suppressive soil (SS) from Chateaurenard. % wilted plants determined 6 weeks after infestation with *F. oxysporum* f.sp. *lini* at different concentrations. (From Alabouvette, C. *et al.* (1985). *In* Parker, C. A. *et al.* pp. 101–6.)

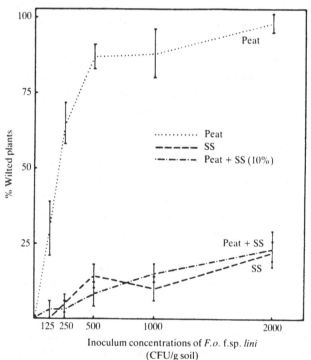

Inoculum concentrations of *F.o.* f.sp. *lini*
(CFU/g soil)

Fig. 5.4. (*a*) Role of the principal fungi occurring naturally in
Chateaurenard soil suppressive to *F. oxysporum* f.sp. *melonis*. (*b*)
Role of three species of *Fusarium* in this soil, applied in various
mixtures and singly. SS = suppressive soil. (From Alabouvette, C. *et
al.* (1979). *In* Schippers, B. & Gams, W. pp. 165–82.)

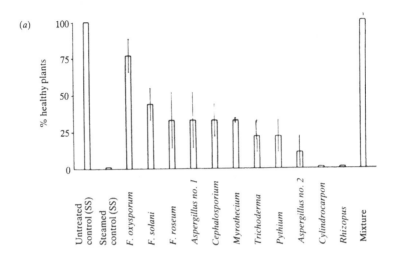

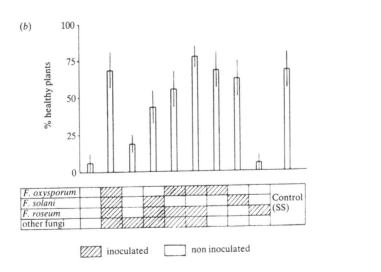

There are soils in Central America that are suppressive to *F. oxysporum* f. sp. *cubense* (banana wilt, Panama disease), but this is attributed mainly to montmorillonite clays in the soils and to alkaline pH, rather than specific microbial antagonism. Such conditions favour the bacteria and reduce growth of fungi so there may still be a microbial component of this suppressiveness.

Other fusaria (*F. oxysporum* f.sp. *vasinfectum*) have reduced severity if the disease is preceded by inoculation with avirulent strains so that the induced resistance, by stimulating phytoalexin production and by vascular occlusions, reduces pathogen invasion.

5.2.2 *Take-all decline and suppressiveness*

This must be the most researched case of a suppressive soil phenomenon in which every possible explanation (and several impossible ones!) has been put forward (Hornby, 1979; Asher & Shipton, 1981). Take-all is caused by the fungus *Gaeumannomyces graminis* and the variety that attacks wheat (var. *tritici*) is the one usually studied. It may cause a seedling kill if it attacks early in the growth, but more usually it causes a root rot, which makes plants susceptible to drought stress, may cause early maturation and reduces yield. In extreme cases there may be no yield at all – although the plant survives it produces no grain. Infected fields may have severe yield reductions, but the average loss over all wheat grown in the UK is perhaps 2% which in 1983 represented £M24: 'take-all is the main root rot and the principal uncontrolled disease of cereals' (Hornby, 1985). It is estimated that in north west USA the take-all losses are 5–10% of the yield. There are no commercially available means of chemical control and no effective resistance in the hosts. As might be expected with a disease which survives on crop residues, it increases if the host is grown repeatedly on the same field and inoculum builds up. Similarly it is favoured by early sowing of winter crops, so there is only a very small gap between the harvest of one crop and the sowing of the next susceptible host. Cultivation, fallow and long rotations without cereals decrease the disease. If monoculture (the same crop year after year) is continued past the first 3 to 5 years of disease build-up it is found that the amount of disease decreases, especially the number of seriously infected plants: this is known as take-all decline and it occurs worldwide, has been studied for 50 years, and is commonly supposed to be caused by the development of a suppressive soil in the prolonged presence of the host and the pathogen. There are claims for much longer times for the development of decline, even up to 70 years of monoculture before the soil becomes suppressive. Decline has all the characteristics of specific suppression (Rovira & Wildermuth, 1981). It can be transferred to conducive soils, though rather large amounts (12.5% w/w) may be needed. The suppression is destroyed by heating, and there are experiments that show that as decline develops there is an increase in the numbers of micro-organisms which are antagonistic to the pathogen in culture and an increase in specific components of the microflora on the roots. For example, in some Australian work the proportion of the pseudomonads which were *P. fluorescens* antagonistic to *G. graminis*, increased. Microbiological effects are clearly important in the development of this suppression. The actual mechanism is unclear, but growth rates of hyphae in take-all decline suppressive soils are less than in conducive ones and this delays the time of maximum numbers of invasive runner hyphae on the roots (Fig. 5.5) and some hyphal lysis was reported. Plants growing in take-all decline soils may still have lesions, but either they do not develop or they spread more slowly.

It has been suggested that take-all decline may be due to a reduction in pathogen virulence with time, and virus infection of the fungus was put forward as the reason for this, but it is not now thought to be important.

Fig. 5.5. Density of the runner hyphae of *G. graminis* on wheat roots. Plants were grown in fumigated soil inoculated with the pathogen (●——●); with the addition of 1% partially suppressive soil (△——△); or with the addition of 1% fully suppressive soil (○——○). (From Wildermuth, G. B. & Rovira, A. D. (1977). Reprinted with permission from *Soil Biology and Biochemistry* **9**, 203–5. © 1977 Pergamon Journals Ltd.)

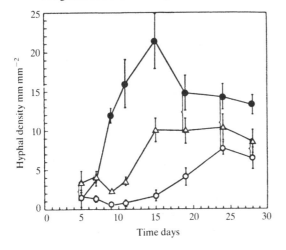

A very large number of potential antagonists to *G. graminis* have been isolated from various soils, including take-all decline soils, and have been tested as biocontrol agents for direct use as inoculants. These include mycophagous amoebae, many bacteria (*Bacillus, Streptomyces, Pseudomonas* – especially *P. fluorescens* and *P. putida*), fungi in general and some closely related to the pathogen such as *Phialophora* spp. and avirulent strains of the pathogen itself (see section 5.5 for the direct use of these inocula). They have been used for reducing inoculum of *G. graminis* and as direct antagonists to the growing mycelium by the production of antibiotics, siderophores and lytic enzymes (see sections 1.3.3 and 1.3.4). The pH in the rhizosphere can affect the take-all fungus and this has been studied especially in connection with different forms of nitrogen available to the plant. Ammonium does not favour take-all and its effect seems to be linked with the presence of *Pseudomonas fluorescens* on the roots, though there could be direct pH effects caused by the plant uptake of ammonium.

It seems very likely that there are many mechanisms, some of them microbiological, for take-all decline (Hornby, 1979) which may operate in different soils, under different climatic conditions or on different continents. It is the very widespread nature of this form of suppressiveness that initially gave a naïve hope for great progress, but ultimately has ended up by producing much literature with apparently conflicting results. To study and understand a particular suppressive soil in a defined region, such as the *Fusarium* suppressive soils of Chateaurenard or of parts of California (see above), is possible. To expect to produce a uniform, worldwide result from experiments conducted under so many different conditions was, with hind-sight, wildly optimistic given the present limited state of our understanding of soil microbiology. However, there have been many individual results, which we will discuss later in this chapter, where potential biological control systems have been or are being developed for take-all. Many of these arose from studies of the decline phenomenon and they show promise for success as biocontrol systems, even if the development of a unifying theory for take-all decline was not helped by the experiments.

Take-all decline is the decrease in disease in the presence of the host and the pathogen, but there is another form of suppressiveness involving take-all which develops with the pathogen but without the host (Gerlagh, 1968). Gerlagh worked on recently reclaimed Dutch polders where take-all occurred in the first wheat crops and on grasses. It probably came from growth on the original reeds, though infection of the newly exposed soil remained a possibility. Reduced disease developed with just the virulent pathogen growing on other hosts. No particular antagonistic organisms seemed to be primarily responsible, but the phenomenon was heat sensitive and transferable.

In addition Gerlagh noted a general suppressiveness of some soils, thought to be partly due to microbial activity, which was an innate property of the soils and did not need the presence of either the host or the pathogen for its development. The general suppressiveness was not easily transferable and was not very heat sensitive.

5.3   Organic amendments

There are many reports of the favourable effects of organic manures and amendments on the health of plants, though they may be anecdotal rather than quantitative. Organic matter may operate in a variety of ways such as improving soil structure or plant nutrition, as well as by affecting disease. We will be concerned only with those organic amendments that have been clearly shown to reduce plant disease by microbiological means such as direct antagonism (antibiotics, lysis, etc., sections 1.3.3 and 1.3.4) or by the induction of fungistasis (section 1.3.5).

Several examples have already been noted where general activity is stimulated by the addition of available carbon sources (section 1.6). The best known of these is the traditional use, commercially and by private gardeners, of organic matter to control *Streptomyces scabies*, potato scab. As it is caused by a bacterium, there is no commercially available, effective chemical control for this disease. Green organic matter incorporated in the planting trench increases general microbial activity which antagonizes the *Streptomyces*. More specific results have also been reported in which *Bacillus subtilis* and saprophytic *Streptomyces* were encouraged by barley, alfalfa or soy meal. The latter was also a substrate for the production of antibiotics against the pathogen.

A general rise in the soil organic matter levels has also been shown to give control of *Phytophthora cinnamomi* rot of avocado in Australia. Extensive investigations of soils suppressive and conducive to the rot were made and the important factor was the amount of organic matter, exchangeable calcium and the general level of fertility. The cultural regime involves adding 10 tonnes ha^{-1} $year^{-1}$ of chicken manure, plus NPK fertilizer and dolomite to correct the pH drift to acidity. In addition to the avocado leaves, legume and maize cover crops are disced into the soil surface to give bulk organic matter. This results in the enhancement of microbial activity during the decay of all the organic matter. Scanning electron microscopy confirmed the increase in bacterial numbers and also the lysis of hyphae and sporangia, probably by *Pseudomonas, Bacillus* and *Streptomyces* which were isolated from the lysed hyphae (Malajczuk, 1979). Various protozoa were also recorded near the lysed hyphae.

Phymatotrichum omnivorum is, as its name suggests, a pathogen of the roots of many plants with over 200 dicotyledons recorded as hosts. It has been studied especially in cotton and alfalfa (lucerne) in the southern USA. It is controlled by rotations with a high proportion of monocotyledonous plants like cereals, but all sorts of organic amendments and ploughing in of green crops help in the control. This gives increases in the general microbial activity, and in particular *Trichoderma* increases, which is correlated with the lysing of the sclerotia in the soil.

Of the many individual organic substances which have been used, chitin is perhaps the best documented. The general thought behind this is that organisms which can degrade chitin will be encouraged in the soil and they might be able to degrade the cell walls of basidiomycete and ascomycete fungi and their various imperfect stages which contain chitin. There is also a great deal of chitin, from shellfish processing industries, looking for a useful and preferably profitable outlet! Its use decreased infection by *Rhizoctonia* and gave a 5- or 6-fold increase in the number of organisms which were antagonistic in culture. Wilt of peas (*Fusarium oxysporum* f. sp. *pisi*) can be reduced by up to 82%, though the time of addition is critical and the substrate must be added some weeks in advance of the sowing of the crop. Over this time (Fig. 5.6) the numbers of fungi and actinomycetes that could be recovered from the soil rose up to 25-fold, the numbers of viable propagules of *F. oxysporum* f. sp. *pisi* was halved in the rhizosphere, and the amount of wilt decreased.

The question of the microbial effects of general composts, especially those used in horticulture as growing media, is more complicated, but it has become important as novel growing media based on peat, composted ground-up bark or municipal waste are used (Hoitink & Fahy, 1986). Traditionally steamed soil, which was more or less sterile, was mixed with sand and peat. Peat, especially that derived from *Sphagnum* or similar mosses, may have a microflora dominated by fungi (*Trichoderma*) and some actinomycetes (*Streptomyces*) which can antagonize *Rhizoctonia, Pythium* and *F. oxysporum* f. sp. *lycopersici*. However, some fen peats only develop suppressiveness to f. sp. *lycopersici* after repeated crops (see section 5.2) and here the microbiological causes are suggested by the loss of suppression on steaming.

There are composts made from municipal waste that has been screened and sorted to remove most of the glass and metal. A nitrogen source is often added as well and this is frequently slurry from sewage treatment plants. Chipped or macerated tree bark is also used as a basis for composts, again with the addition of nitrogen. Many potential animal and plant pathogens are killed during the heating of the compost as it decomposes, but this can also kill potential antagonists and care is needed in the size and design of the heap of material to reduce the natural heating or to turn the heap so that organisms near the edges which have survived the heating can recolonize the interior as the decomposition passes its main peak of activity. In general, municipal waste compost is less suppressive than bark composts, probably because the former has been allowed to heat up more in order to kill possible faecal pathogens in the added slurry. There are clear microbiological differences between the two composts with antagonistic bacteria (*Pseudomonas, Bacillus, Enterobacter cloacae*) and fungi such as *Trichoderma* and *Gliocladium virens* present in those that allow development of less disease. It is also possible to add back suitable

Fig. 5.6. Effect of chitin on rhizosphere and non-rhizosphere micro-organisms of pea grown in wilt infested soil in the field. Chitin at concentrations (g/plot) of: 0 = ○---○; 35 = △ — △; 70 = x——x; 140 = ●——●; 280 = ○——○. (*a*) Actinomycetes. (*b*) Average numbers of *F. oxysporum* f.sp. *pisi* in the rhizosphere ●——● and non-rhizosphere ○---○ soil treated with chitin. (From Khalifa, O. (1965). *Annals of Applied Biology* **56**, 129–37.)

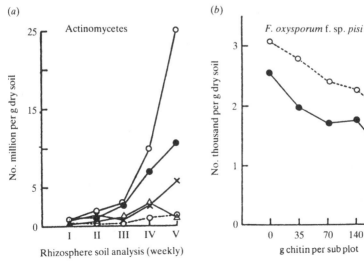

organisms, preferably several different ones, to make the conducive, heated composts from the centre of the heaps suppressive to *Pythium* and *Fusarium* wilt.

Composted municipal waste, used as an organic manure, has been reported to reduce *Sclerotinia minor* on lettuce and this is correlated with a rise in general microbial activity as measured by dehydrogenase activity in the soil. There is no evidence of specific organisms being involved.

5.4   Mycorrhizae and root disease

Ectomycorrhizae are associations between fungi, mostly basidiomycetes, and the roots of some temperate forest trees in which the fungus forms a sheath over the root and hyphae spread out into the soil. Hyphae may penetrate between the cells of the epidermis and outer cortex, but they stay outside the cell walls. They have been particularly studied in relation to nutrient uptake, but they also affect root disease. As they completely surround the root they change the quantity and quality of the exudates that reach the soil and they have a different rhizosphere population from uninfected roots (Campbell, 1985). In those combinations of hosts and symbiont, in which the sheath is continuous and quite thick, the fungus can form a physical barrier to infection, preventing pathogens from reaching the root surface, as in the case of the mycorrhizal fungus *Pisolithus tinctorius* excluding *Phytophthora cinnamomi* from the root of eucalyptus. This pathogen has caused serious dieback of the native jarrah forests of Australia by killing the young feeding and water absorption roots of the trees. Other mycorrhizal fungi such as *Leucopaxillus cerealis*, *Laccaria laccata*, *Lactarius deliciosus* and *Suillus luteus* are known to produce antibiotics which are effective, in plate tests, against *P. cinnamomi* and many other potential root pathogens.

Similar conclusions were drawn from a study of pine roots that were mycorrhizal with *P. tinctorius* or *Thelephora terrestris*, neither of which produce antibiotics. The infection by *P. cinnamomi* was reduced to less than 10% of that of non-mycorrhizal roots, and where infection did occur it was through the root meristems which are not protected by the mycorrhizal sheath.

Vesicular arbuscular mycorrhizae (VAM) are associations between the roots of many species of plants, including most agricultural crops, and phycomycete fungi. The fungal hyphae penetrate the cell walls, though not the plasmalemma, and form highly branched structures called arbuscules. These are transitory, lasting perhaps 10 days, and are thought to be the sites of mineral absorption into the plant. There are also thick-walled vesicles between the cells and spores are produced at the root surface or in the soil. In this form of mycorrhiza there is no sheath as described above for the ectomycorrhizae. The effects of VAM on disease are quite complicated (Bagyaraj, 1984), but usually beneficial (Table 5.1), though they may encourage some diseases, such as *Phytophthora* root rot of soybean, but this latter activity is unusual. Frequently they seem to have no effect, or the action may be indirect such as increased lignification in the root preventing penetration by some *Fusarium* species. In general the VAM need to be pre-inoculated, or better still to have already infected the root to give disease control.

The increased phosphorus levels in the root, for which mycorrhizae have mostly been studied, can themselves decrease root exudation and this has been given as a reason for less stimulation of spore germination or growth of pathogens in the rhizosphere. Less infection by *G. graminis* (take-all) at low inoculum levels is attributed to less growth in the

Table 5.1 *Effects of VA mycorrhizae on soil-borne disease caused by fungi*

Pathogen	Host	Effects in mycorrhizal plants
Olpidium brassicae	tobacco lettuce	reduction of infection
Pythium ultimum	soybean	none
Pythium ultimum	poinsettia	reduced stunting
Phytophthora megasperma	soybean	fewer plants killed
Phytophthora palmivora	papaya	none
Phytophthora parasitica	citrus	reduction of damage
Rhizoctonia solani	poinsettia	reduced stunting
Thielaviopsis basicola	tobacco	⎫ less stunting, inhibition
Thielaviopsis basicola	alfalfa	⎬ of chlamydospore
Thielaviopsis basicola	cotton	⎭ production
Cylindrocarpon destructans	strawberry	⎫
Cylindrocladium scoparium	yellow poplar	⎪ less stunting, reduction
Fusarium oxysporum	tomato	⎬ of infection
Fusarium oxysporum	cucumber	⎪
Phoma terrestris	onion	⎭

Compiled by Schönbeck, F., 1979. (See Schippers, B. & Gams, W., pp. 271–80.)

rhizosphere depleted of nutrients (Table 5.2). There is also a correlation with phosphorus level *per se*: take-all is known to be serious in phosphorus deficient plants, and simply adding phosphate fertilizer is effective in reducing the disease in these situations (Table 5.2). VAM do the same thing as the fertilizer by increasing phosphorus levels, and it is probably nothing in particular about the VAM that is giving disease control, just the effect on plant nutrition.

There are similar rather devious effects reported for the interaction of VAM with *Thielaviopsis basicola* rot of cotton roots. The VAM do not seem to decrease infection, though they do give better growth and dry weight of the plant, probably again a plant nutritional effect. However, there are fewer chlamydospores produced by the pathogen and so less infection in subsequent crops. Amino acids, especially arginine, inhibit chlamydospore formation by the pathogen and roots infected with VAM have 50% more amino acids than uninfected roots.

Table 5.2 *Influence of soil phosphorus and VA mycorrhizae on the severity of take-all disease of wheat*

Treatment	VAM formation (%)[1] NM[3]	VAM[3]	Roots lesioned (%)[2] NM[3]	VAM[3]	Disease rating (0–4) NM[3]	VAM[3]
0 P[4] – 0[5]	...	92 a[6]	0 a[6]	0 a[6]	0 a[6]	0 a[6]
0 P – 0.1	...	94 a	46 b	33 b*	3.2 b	2.3 b*
0 P – 0.5	...	94 a	53 c	44 c	3.4 b	3.0 c
50 P – 0	...	14 b	0 a	0 a	0 a	0 a
50 P – 0.1	...	11 b	30 d	31 b	2.2 c	2.2 b
50 P – 0.5	...	12 b	33 d	30 b	2.4 c	2.3 b

[1] Percentage of root length with mycorrhizal structures present.
[2] For disease assessment, roots were rated visually on a 4–0 scale, and the percentage dry weight of lesioned root tissue of the total root dry weight determined.
[3] Means for VAM (mycorrhizal with *Glomus fasciculatus*) treatments followed by an asterisk (*) are significantly different from the respective NM (nonmycorrhizal) mean, $P = 0.05$.
[4] Concentrations (0 and 50 µg P/g soil) of P added to soil as superphosphate.
[5] Inoculum level (grams of inoculum per gram of dry soil) of *G. graminis* var. *tritici*.
[6] Values are the mean of 10 replications. Column means followed by the same letter are not significantly different according to Duncan's multiple range test, $P = 0.05$.
From Graham, J. H. & Menge, J. A. (1982). *Phytopathology* **72**, 95–8.

This again is an indirect effect on host physiology. There seems to be no clear evidence, for any combination of host and phycomycete symbiont, that VAM produce antibiotics or otherwise directly inhibit pathogens.

The whole subject of mycorrhizal effects on pathogens is worthy of further study. It is probable that mycorrhizal inoculants will eventually be used commercially, with or without biological control, for their beneficial nutrient effects. This is already true on a limited scale for ectomycorrhizae, but at the moment VAM fungi cannot be grown in commercial quantities in culture. An additional benefit from the control of pathogens, if it could be shown to occur consistently, would increase the possibility of developing inoculants that were economically viable and seen by farmers and foresters as worth the trouble to use.

5.5   Cross-protection and induced resistance

Cross-protection is the prevention of disease by the use of an organism similar to the real pathogen. Induced resistance is a particular type of this phenomenon in which the host defence mechanisms recognize and respond to the harmless mimic and are then ready ahead of the real threat posed by the later arrival of the pathogen. Cross-protection may therefore be by induced resistance, or by many other methods of antagonism.

The organism used may be an avirulent strain of the pathogen (see the discussion of *Endothia* on stems, section 4.2.3) or a different forma specialis (section 3.6.2, on leaves), or even a different but related species. In soil, the cross-protection by *Agrobacterium* strain 84 has already been described (section 4.4); an avirulent strain is used and it operates by the production of a bacteriocin not by induced resistance. There are reports of the use of strains of *F. oxysporum* f. sp. *melonis* but these may not be entirely avirulent, so there is a risk of introducing the disease itself, and anyway it is probable that the mechanism is competition for entry sites on the host – the protective strain entering but causing less disease than the real pathogen – so again this is cross-protection but not induced resistance.

There are very few well documented cases of induced resistance operating against soil-borne pathogens, and most of the ones that there are involve wilt diseases. Tomatoes can be protected from *F. oxysporum* f. sp. *lycopersici* by dipping the roots in a suspension of f. sp. *dianthi* some days before likely exposure, but protection only lasts a few weeks. Protection for at least 3 months was provided by spraying the roots of cotton at transplanting with a mildly pathogenic strain (SS–4) of the wilt *Verticillium albo-atrum* (virulent strain T–1). This reduced the amount of severe infection by 94% (Table 5.3). Similarly the inoculation of mint (*Mentha* spp.) with the mildly pathogenic *V. nigrescens* greatly reduced the serious symptoms and death caused by *V. dahliae*. Furthermore there was a reduction in the viable propagules of the pathogen left in the stems (Fig. 5.7) so the disease might also be reduced in the longer term.

There is at least an element of induced resistance in the use of some fungi against *G. graminis* var. *tritici* on wheat. Some *Phialophora* spp. are known to be the imperfect stages of *Gaeumannomyces*, though there are some which have no known perfect stage. *Phialophora* and *G. graminis* var. *graminis* grow on grass roots and can also be found on wheat where they occupy a very similar niche to the pathogen. They invade the root cortex but not the stele, and are halted by lignification of the cortex in general, by the production of lignitubers (localized thickening of the wall at the point of attempted entry by a fungus), and by an increase in the lignification and suberization of the endodermis and the stele. The root cells with chemically changed walls and extra thickening are less susceptible to invasion by *G. graminis*. This is the stimulation of host defences that constitutes induced resistance. There is also a component of competition for available space for infection and spread, because root cortex occupied by *G. graminis* var. *graminis* or *Phialophora graminicola* cannot be invaded by *G. graminis* var. *tritici*:

Table 5.3 *Cross-protection with a reduced virulence strain (SS–4) of* Verticillum albo-atrum *in a field of cotton infested with a virulent strain (T–1) of the pathogen*

| Treatment | No. of plants with symptoms[a] | | |
	None[b]	Mild[c]	Lethal[d]
Nonsterile field soil[e]	2	0	16
Nonsterile field soil + 10^5 propagules of SS–4	0	17	1
Sterile field soil + 10^5 propagules of SS–4	4	14	0
Sterile field soil	18	0	0

[a] Final readings taken 3 months after planting.
[b] Average escape where mixture of strains was not used was 8%.
[c] SS–4 symptoms.
[d] T–1 symptoms.
[e] Nonsterile field soil naturally infested with T–1 strain was estimated to have 5×10^3 viable propagules g^{-1} of dry soil.
From Schnathorst, W. C. & Mathre, D. E. (1966). *Phytopathology* **56**, 1204–9.

the initial colonizers are good at defending a captured resource (section 1.2). In Europe this cross-protection increases the yield of wheat growing in soil infected by the pathogen (Table 5.4) and it works best when the saprotroph has more inoculum than the pathogen. *Phialophora* grows on grass roots, and grass leys in the rotation before wheat leave some *Phialophora* to give protection for the wheat, but the *Phialophora* does not persist. These fungi do not seem to work in the USA and either do not occur in Australia or are uncommon (*G. graminis* var. *graminis*). The only effective *Phialophora* that occurs in Australia is *P. hoffmanii*. If introduced into wheat in Australia the *P. graminicola* does not work as it does in Europe, and the *G. graminis* var. *graminis* and *Phialophora* sp. (which is probably *G. graminis* var. *graminis* as well) are not very effective (Table 5.5), though there are slight yield increases over the diseased control.

This *Phialophora* story is very confused, with different results in different countries and this is not helped by taxonomic confusion so that it is not always clear whether the same organisms are being used, never mind effective strains of the same species. It is a common story for biological control of soil diseases, and take-all in particular: there seems to be something useful going on, but it cannot be really defined, quantified and repeated successfully.

Fig. 5.7. Propagules of *Verticillium* in stems of cross-protected and non-protected peppermint, inoculated with the avirulent *V. nigrescens* 7 days before the virulent *V. dahliae*. (From Melouk, K. & Horner, C. E. (1975). *Phytopathology* **65**, 767–9.)

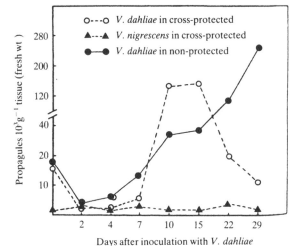

Table 5.4 *The effect of inoculation of wheat with* G. graminis *var.* graminis *on the take-all disease caused by* G. graminis *var.* tritici. *Grain yield, 1000 grain weight (a measure of seed size) and plant height are shown*

Treatment	Mean grain yield (g/plot)	Mean plant height (cm)	TGW[a] (g)
Healthy control	487.43	106	41.4
G.g. tritici alone	153.22 (31.4)*	85	38.9
G.g. graminis alone	524.44 (107.6)	110	43.0
tritici + *graminis* (1 : 1)	296.88 (60.9)	100	42.9
tritici + *graminis* (3 : 1)	272.75 (56.0)	90	39.6
tritici + *graminis* (1 : 3)	412.21 (84.6)	106	41.2
L.S.D. ($P \leqslant 0.05$)	78.77		2.3

[a] TGW – Thousand grain weight.
* Figures in parentheses indicates the yield as % of that of healthy control.
From Speakman, J. B. (1984). *Phytopathologische Zeitschrift* **109**, 188–91.

Table 5.5 *The effect of various* Gaeumannomyces *and* Phialophora *strains (avirulent) giving cross protection, including induced resistance, against take-all of wheat*

Treatment	Mean grain yield (kg/plot)	% wheat plants with roots colonized by	
		G. g. *tritici*	Avirulent fungus
Cross-protected (*G. g. graminis*)	2.00 (45.1)*	47.5	41.8 ± 3.6
Cross-protected (*Phialophora* sp.)	2.03 (45.7)	50.7	26.7 ± 2.9
Cross-protected (*P. r. graminicola*)	1.57 (35.4)	49.3	24.3 ± 1.7
Unprotected	1.40 (31.5)	54.2	0.0
G. g. graminis alone	4.33	0.0	60.8 ± 2.1
Phialophora sp. alone	4.55	0.0	28.8 ± 3.4
L.S.D. ($P = 0.05$)	0.27	13.7	

* Figure in parentheses indicates the yield as a % of the mean yield of treatments inoculated with avirulent fungi.
From Wong, P. T. W. & Southwell, R. J. (1980). *Annals of Applied Biology* **94**, 41–9.

5.6   Direct inoculation of fungal and bacterial antagonists

This is the consideration of the biological control of root diseases by the direct inoculation of soil or seeds with fungal and bacterial antagonists which operate by inhibiting or killing the pathogen. The use of plant growth promoting rhizobacteria and amoebae is discussed later (sections 5.8 and 5.7, resp.).

The problems with the direct inoculation of soil have already been discussed (section 5.1), but briefly the organisms used must be normal soil inhabitants which will grow and spread on roots or in soil and which will survive for a sufficient period of time. The greatest difficulty is to get repeatable results from year to year and from site to site. This may be a variability in the quality or purity of the inoculum or the effects of different climatic and edaphic factors.

One must also distinguish between the failure of the trial because the proposed antagonist did not operate as expected, and the case in which there was no difference between the diseased controls and the treatment because the disease was present at such low levels that the organism was not seriously tested for its ability to control the disease. In the latter case the antagonist has not failed, the experimenter has failed to select a suitable trial site! With some diseases it is, in fact, quite difficult to be sure that there will be disease in a given field, or worse a given part of the field. Take-all is well known for its patchy distribution and variability from year to year. One solution to this may be to inoculate the pathogen artificially, perhaps even after fumigation, and this will usually give a more even distribution of disease and a 'better' trial with less variability. It is highly artificial, but may be useful for testing potential antagonists. However, sooner or later the biological control will have to be shown to work, i.e. give disease control or useful yield increases, under natural disease conditions.

There are many reports in the literature that describe the control, usually under laboratory or glasshouse conditions, of a vast range of diseases. However, these are only rarely followed up to take account of the various problems outlined above, so here we will consider in detail only those potential control agents which have been seriously tested over several years, in different soils under realistic agricultural conditions. This reduces the list of antagonists active against root diseases to three main groups, the fungi (principally *Trichoderma*), *Bacillus* and the pseudomonads.

5.6.1   Trichoderma

Trichoderma has been used against the wilt diseases of tomato, melon and cotton, and *Fusarium culmorum* on wheat (Table 5.6). *Trichoderma harzianum* gave 60 to 83% control of these *Fusarium* diseases in naturally infected field soil. Inoculation with *Trichoderma* was done either to the seeds or as a bran mixture incorporated into the transplanting compost. Disease did eventually develop, but at a much slower rate with the treatment and, in the case of melons (Fig. 5.8), was significantly less after 13 weeks. Furthermore the control persisted in the soil and gave measurable decreases in disease for three successive plantings of melons. Control of *Fusarium* wilt of chrysanthemums is also known to occur with *Trichoderma*.

There is no clear information on the mode of action in these studies with *Fusarium*. With *Verticillium albo-atrum* causing wilt of tomato there is, however clear evidence of antibiosis. Spore suspensions and the culture filtrates of *T. viride* (amongst other fungi) significantly increased the yield of fruit over the diseased control, and beyond even the healthy control in some cases. It was suggested that the filtrates inhibited both the germination and the subsequent growth of the *V. albo-atrum*.

Various species of *Trichoderma* have been extensively tested against damping off diseases of seedlings (*Rhizoctonia*, *Pythium*, etc., see section 7.4). This overlaps with their use against *Sclerotium rolfsii* which may cause seedling diseases but also a root rot of more mature plants in tropical and sub-tropical regions where soil temperatures exceed 25 °C (in the USA it is known as southern root rot). It attacks many hosts

Table 5.6 *Biological control of* Fusariam *spp. by* Trichoderma harzianum

Pathogen	Crop	Disease[a] reduction (%)
F. oxysporum f. sp. *vasinfectum*	cotton	80
F. oxysporum f. sp. *melonis*	melon	60
F. oxysporum f. sp. *radicis-lycopersici*	tomato	80
F. culmorum	wheat	83

[a] Percentages of disease reduction (DR) were calculated according to the following formula: DR = $[1 - DT/DC] \times 100$, where DC and DT are percentages of disease in control and treatment, respectively.
From Sivan, A. & Chet, I. (1986). In *Microbial communities in soil*, ed. V. Jensen *et al.*, pp. 89–95. London: Elsevier Applied Science Publishers.

including groundnut (peanut), cotton, beans, potatoes and tomatoes and it survives unfavourable periods by forming sclerotia in the soil. Strains of *T. harzianum* have been isolated that possess β 1–3 glucanases, chitinases and sometimes proteinases that enable them to parasitize the hyphae and sclerotia of the pathogen, invading the cells and causing lysis. Good control of *S. rolfsii* has been obtained in field conditions by several groups of workers in different parts of the world. The *T. harzianum* is grown on autoclaved seed or bran and then mixed with the surface soil (Chet & Henis, 1985). Other sclerotial fungi, such as *Sclerotium cepivorum*, the cause of white rot of onion, have also been controlled on a field scale by *Trichoderma*, and again the mechanism is mycoparasitism.

The sclerotia of *Rhizoctonia solani* are important in the persistence of rice sheath blight which, though a leaf disease, is controlled during the survival stage in the soil. *T. harzianum* destroys the straw on which the pathogen survives and also parasitizes the sclerotia, so reducing the viable *R. solani* in straw from 100% occurrence in freshly incorporated

Fig. 5.8. Biological control of *Fusarium oxysporum* f.sp. *melonis* by *T. harzianum* under field conditions. (*a*) *Trichoderma* applied as a seed coat (●——●) in a naturally infested, untreated soil (○——○). (*b*) *Trichoderma* mixed with rooting medium (▲ — ▲) compared with the untreated soil (△ — △). (From Sivan, A. & Chet, I. (1986). In *Microbial communities in soil*, ed. V. Jensen *et al.* pp. 89–95. London: Elsevier Applied Science Publishers.)

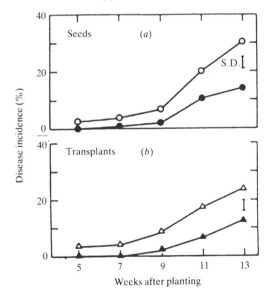

straw to 20% after 2 weeks and causing complete removal of the pathogen by 16 weeks (Fig. 5.9). Various other biocontrol agents (e.g. *Pseudomonas*) are also known to operate against rice sheath blight.

There are two fungi that are especially known to parasitize sclerotia and they are *Coniothyrium minitans* and *Sporidesmium sclerotivorum*. *C. minitans* was first suggested as a biocontrol agent in 1947 and has been regularly recommended ever since. It occurs worldwide, always associated with sclerotia of a variety of plant pathogens such as *Sclerotinia sclerotiorum, Sclerotium trifoliorum, S. cepivorum, Botrytis cinerea, B. fabae, B. narcissicola* and *Claviceps purpurea* (Ayers & Adams, 1981). *C. minitans* produces several enzymes which allow it to cause lysis and it can kill up to 99% of the sclerotia in 11 weeks, greatly reducing the inoculum potential of the pathogen. There are correspond-

Fig. 5.9. The effect of *T. harzianum* on *Rhizoctonia solani* causing sheath blight in rice. (*a*) *Trichoderma* (———) causes pigmentation of the straw, indicating decomposition; this is more rapid than with *Rhizoctonia* alone (—·—·) and the mixture is intermediate (– – – –). (*b*) *R. solani* is recovered much less frequently from the straw decayed by *Trichoderma* (– – – –), only 20% of the straw has the pathogen after 2 weeks and it is not possible to recover it after 14 weeks. *R. solani* alone (—·—·) last much longer. (From Mew, T. W. & Rosales, A. M. (1985). *In* Parker, C. A. *et al.* pp. 117–23.)

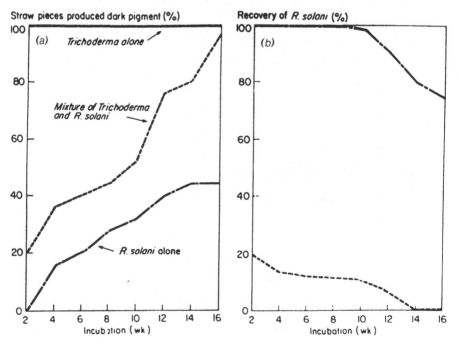

ing reductions in the disease and increases in yield. *Sporidesmium sclerotivorum* conidia are stimulated to germinate by exudates from the pathogen sclerotia which are then invaded and lysed. Again this greatly reduces the inoculum potential (Fig. 5.10) while giving up to 83% reduction in disease over periods of up to a year against *Sclerotinia minor* on lettuce. There is a problem with *Sporidesmium sclerotivorum* in producing spores in culture for the inoculum, but this should be soluble. These two specialized fungi, known almost exclusively from sclerotia, and the *Trichoderma* isolates noted above, are good examples of the reduction in pathogen inoculum potential by the direct addition of an antagonist, rather than attacking the colonizing hyphae near or on the host. Reduction in inoculum is particularly useful when the pathogen has long-lived sclerotia in soil that are difficult to reach with fungicides, short of using soil fumigation.

Fig. 5.10. Infection (●—●) and survival of sclerotia of the pathogen *Sclerotinia minor* in loam treated with a culture of *Sporidesmium sclerotivorum* (○---○). The survival of the pathogen in untreated soil is also shown (▲—▲). (From Ayers, W. A. & Adams, P. B. (1981). In *Biological control in crop production*. ed. G. C. Papavizas, pp. 91–103. Granada: Allanheld, Osmum Publishers.)

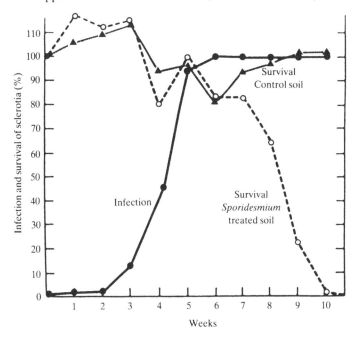

5.6.2   Bacillus

This bacterium occurs regularly amongst those that are proposed for the biological control of many root diseases and has various advantages, especially that as it forms spores it is easy to prepare inocula which have a very long shelf-life, and it persists in the soil. The corollary of this is that, though it may be present, it may be dormant. In various studies *Bacillus* has been selected in initial screening processes but is then not retained in later work. In general it is a less good colonizer of roots than *Pseudomonas* and less versatile nutritionally (see below). It has been shown to give very good control on occasion (Fig. 5.11) with a doubling of wheat yield in this case compared with plants infected with natural take-all. *B. subtilis* and *B. pumilus* have also been used to protect wheat from *Rhizoctonia* (Table 5.7). There is much less root discoloration in the presence of the bacteria and this decrease in disease was also reflected in increased yield.

Fig. 5.11. Yield (a) and 1000 grain weight (b) for spring wheat in soil naturally infected with take all and inoculated with *Bacillus pumilus*. ■ uninoculated control; ℕ inoculated at sowing; ▤ inoculated 6 weeks after sowing; ▥ inoculated at sowing and 6 weeks after sowing; ▦ double strength inoculum at sowing; □ double strength inoculum at both times. (From Capper, A. L. & Campbell, R. (1986). *Journal of Applied Bacteriology* **60**, 155–60.)

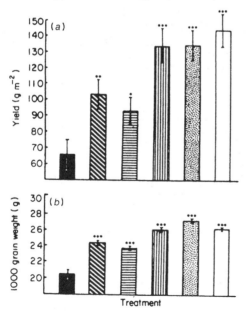

The main disadvantage of *Bacillus* seems to be that it is specially prone to give results that are variable. The *Bacillus pumilus* used in Fig. 5.11 has produced disease control in other years and on other sites, but set against this is that in the same year it produced no effect in other fields close by and it has produced only transient disease control, in

Table 5.7 *The effect of antagonist treatments on yield and on symptoms caused by* R. solani *on wheat in natural soil inoculated with the pathogen and a variety of bacteria*

Seed treatment	Mean grain weight (g)		
	Without *R. solani*	With *R. solani*	Combined mean
Nil	21.93	20.98	21.45
S. griseus (2–A24)	26.43	27.68	27.06
B. sub. (1–B80)	21.92	27.41	24.66
B. sub. (1–B77)	25.04	22.85	23.94
B. sub. (1–B68)	24.96	24.23	24.60
B. sub. (1–B3)	24.86	19.93	22.39
Bacillus sp. (1–B8ii)	22.92	20.77	21.84
B. pum. (1–B84)	25.63	22.37	23.99
LSD *P* = 0.05	3.06		2.16
LSD *P* = 0.01	4.03		2.85

Seed treatment	Unpasteurized soil		
	Plant height	Root length	Discol. root length
Nil	298	112	0
Nil + *R. solani*	229	90	26
S. griseus 2–A24 + *R. solani*	174	85	9
B. sub. 1–B80 + *R. solani*	191	97	7
B. sub. 1–B77 + *R. solani*	232	115	12
B. sub. 1–B68 + *R. solani*	222	83	9
B. sub. 1–B3 + *R. solani*	165	80	9
Bacillus sp. 1–B8ii + *R. solani*	217	103	17
B. pum. 1–B84 + *R. solani*	124	65	15
LSD *P* = 0.05	78	31	10
LSD *P* = 0.01	104	41	13

Means in millimetres.
From Merriman, P. R. *et al.* (1974). *Australian Journal of Agricultural Research* **25**, 213–18.

some years, which occurred early in the season and was not reflected in the final yield of the crop. Even mean yield increases as high as 10 or 12% can be statistically insignificant because of variability within the experiment. Similarly the results of Merriman (Table 5.7) have been difficult to repeat and have not gone on to produce a commercially useful product. There are many other examples of this: the use of *B. subtilis* gave good control of white rot of onion (*Sclerotium cepivorum*), significantly reducing the infection to half the control level, but in this case even doubling the yield did not give statistically significant results because of the variability in response within the experiment.

5.6.3   Pseudomonas

This genus is undoubtedly considered by most workers to be the one to select as a biocontrol agent for soil-borne diseases, especially the two fluorescent species or groups, *P. fluorescens* and *P. putida*. They are easy to isolate and grow in the laboratory and they are nutritionally very versatile. They can even be identified relatively easily, which is more than can be said of most bacteria from soil! They are normal inhabitants of the soil and especially of the root surfaces of plants, so they grow and colonize very well when introduced artificially (section 5.1, Fig. 5.1). Their generation time can be as low as 5.2 h (compared with 39 h for *Bacillus* above). Furthermore they are known to produce a variety of antibiotics and siderophores, some at least of which are active in soil. They have been implicated in many cases of natural biological control, in connection with plant growth promotion (section 5.8) and suppressive soils (section 5.2.2). In short they are the main hope for the future biocontrol of soil-borne diseases, and various strains and their secondary metabolites have got as far as to be patented, and have now been marketed commercially for root rot of cotton.

As examples of the direct use of *Pseudomonas* as an antagonist we will take the *Fusarium* wilt studied by R. Baker and many co-workers, and the control of take-all studied by Cook and his co-workers. Both of these have been considered before in connection with suppressive soils and the work with the direct addition of antagonists has in both cases grown out of the study of suppression *per se*.

Fusarium oxysporum f. sp. *cucumerinum* and f. sp. *lini*, which attack cucumbers and flax, produce chlamydospores as a survival mechanism. These require ferric iron for germination and germ tube growth at concentrations greater than 10^{-19} M. The availability of iron can be controlled experimentally by the use, in culture and in soil, of different chelating compounds with different stability constants. *P. putida* produces a siderophore with a very high affinity for iron and to complicate the matter a little more, the host plant and the pathogen also produce siderophores. There is, therefore, intense competition for iron in the rhizosphere and if the pathogen does not get enough it neither germinates nor grows, and the disease decreases. Fig. 5.12 summarizes the relative affinities of the different chelating agents and also stresses that overall the reaction is determined by the pH which controls the equilibrium of the dissociation of ferric hydroxide to free ferric ions: at high pH most of the iron is as the insoluble hydroxide and has very low availability and the disease is not important. At acid pH (<6) there are free hydrogen ions and hence free ferric ions, and the disease is serious. The greatest affinity for iron is shown by the *P. putida* siderophore which can out-compete all the others. The host-produced siderophore can remove iron from all except the bacterium, but in the presence of the bacterium there may be induced iron deficiency in some plants. The EDDHA (ethylenediaminedi-o-hydroxyphenyl acetic acid) can also remove iron from the pathogen. All these will, therefore, tend to induce iron shortage in the pathogen and reduce the level of disease. *Fusarium* itself can get iron from the EDTA (ethylenediaminetetraacetic acid). There is a good correlation between the ability of the antagonistic strains to produce siderophores in culture and the ability to reduce chlamydospore germination ($r = 0.99$) or to inhibit the disease in the field ($r = 0.7$–0.9 depending on the f. sp.).

The real test of this proposed system is to monitor the disease in soil under different iron availability conditions, with and without the

Fig. 5.12. The affinity for Fe^{3+} of the siderophores of *P. putida* and *F. oxysporum* in relation to artificial chelating agents and iron availability in the environment. (Based on Baker, R. *et al.* (1986). *In* Swinburne, T. R. pp. 77–84.)

$$3H^+ + Fe(OH)_3 \rightleftharpoons Fe^{3+} + 3H_2O$$

High stability constants   ⟶   lower stability constants

P. putida >> Host   > EDDHA >> *Fusarium*   > EDTA
siderophore   siderophore   siderophore

Iron not available to   Iron available to *Fusarium*
Fusarium unless pH acid   unless pH very alkaline so
so that the system is   that $[Fe^{3+}]$ is very low
Fe^{3+} saturated

bacterium (Fig. 5.13). In the presence of EDTA, with or without the *P. putida* there is serious disease because the system is flooded with available iron, the siderophore of the antagonist is saturated and the pathogen still gets iron. Notice that the EDTA has made the disease worse than usual (the Control columns) by supplying extra iron, suggesting that the disease is normally slightly iron limited. When no artificial iron chelators are added (Control) the disease is serious unless there is the antagonist: the antagonist's siderophore system works and controls the pathogen. EDDHA also controls the disease by depriving the pathogen of iron and reducing Fe^{3+} levels so that the *Pseudomonas* works even better by mopping up the remaining few ions. The final check is the FeEDDHA that shows that it is not the toxicity of the chelator which causes the effect, but the ability to bind iron more strongly than the pathogen. The interaction and cumulative effect between the antagonist and the EDDHA has been further demonstrated (Fig. 5.14). Increasing concentrations of both give increasing levels of control as the iron is sequestered: the greatest control is obtained with the highest numbers of *P. putida* and the greatest concentration of EDDHA.

Fig. 5.13. Mean *Fusarium* wilt incidence when iron chelators were introduced into conducive soil infested with *Fusarium oxysporum* f.sp. *lini* with or without *P. putida*. (From Scher, F. M. & Baker, R. (1982). *Phytopathology* **72**, 1567–73.)

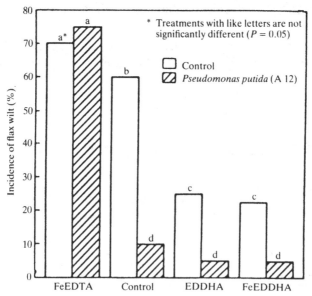

Work with a different strain of *Pseudomonas* (strain B–10) by Kloepper and co-workers at the University of California has given similar results to those of Baker at the University of Colorado and extended it by using not only the bacterium but its purified siderophore, pseudobactin, to produce the control of *F. oxysporum* f. sp. *lini*. Kloepper also showed control of take-all infection with the same bacterium and its siderophore, which leads us onto the next part of the story!

The work on take-all with *Pseudomonas* by Cook's group has been summarized by Weller (1985) and has been mentioned in connection with take-all decline (section 5.2.2). Isolates of *P. fluorescens* from decline soils can be applied as seed coats and inoculated into fields suffering serious take-all where they give 10–27% yield increases compared with the untreated, infected controls. There is now evidence from this, and other groups in Australia, Holland and Great Britain, that *Pseudomonas* can control take-all by both antibiotic production and by the use of siderophores (Table 5.8). The interaction between these modes of action can be complicated. Firstly, there are some *Pseudomonas* strains that give control which do not produce siderophores and some which have no known mode of action, producing neither siderophores nor antibiotics so far as can be determined. The strain 2–79

Fig. 5.14. Effect of increasing levels of FeEDDHA and *P. putida* strain A-12, on cucumber wilt in soil. (From Scher, F. M. (1986). *In* Swinburne, T. R. pp. 109–17.)

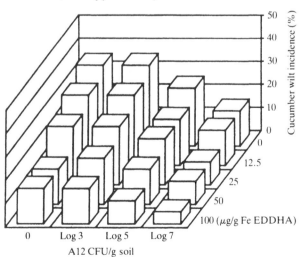

in Table 5.8 produces a siderophore with similar reactions to those described above for EDDHA, EDTA, etc., though there is still about half the inhibitory activity even in the presence of excess iron because it also produces the phenazine antibiotic with a wide range of activity against *Pythium, Rhizoctonia* and *Fusarium culmorum* as well as *G. graminis.* It is thought that the siderophore may be important in early colonization when the take-all fungus competes for iron, but the antibiotic, as a secondary metabolite, may be produced later and inhibit growth in the lesions and the stele (Weller & Cook, 1986). The antibiotic itself can be produced in quantity in culture and is effective as a chemical seed treatment.

There have been many possible biocontrol agents for take-all over the years, and some have even been patented, but they have come to nothing as yet. Past experience suggests that it is unwise to be too hopeful about a control measure for this disease, but there are several groups around the world that have made good progress recently and if something is not produced soon, for all the work that has been done, then the outlook for biocontrol in soil in general is not good. Take-all is in many ways a test case for biocontrol; the disease has been worked on for decades, money has been invested and research is now extensive on possible biocontrol measures. Promising results have been obtained, but if a commercially exploitable control is not available soon then serious doubts will be raised about the ability of biocontrol to work on crops of major world importance under realistic agricultural conditions.

Table 5.8 *The influence of seed treatments with* P. fluorescens *strains 2–79 and 13–79 on take-all of wheat (caused by* G. graminis *var.* tritici = Ggt) *growing in field soil*

Seed treatment	Ggt added	Plants infected	Plant height (cm)	Heads	Yield (g)
1980–1981 Winter wheat plot					
2–79 + 13–79	+	29 b	24 b	287 b	221 b
Check	+	39 c	22 c	255 c	189 c
Check	−	0 a	36 a	429 a	442 a
Probability		0.05	0.05	0.1	0.1
2–79	+	30 b	24 b	296 b	222 b
Check	+	39 c	22 c	265 b	203 b
Check	−	0a	38 a	446 a	465 a
Probability		0.05	0.05	0.05	0.05

From Weller, D. & Cook, R. J. (1983). *Phytopathology* **73**, 463–9.

5.7   Direct inoculation of amoebae

The use of mycophagous soil amoebae (Old & Chakraborty, 1986) as biocontrol agents has always been intuitively attractive. They are so obviously doing something when they make holes in spores or hyphae that you feel they must be reducing the amount of the pathogen.

Some soil amoebae (*Arachnula, Vampyrella, Theratromyxa*) use fungi for food. They may ingest yeasts and small spores whole but they are usually associated with holes, 2–6 µm in diameter, bored in the walls, especially of pigmented fungal structures (Fig. 5.15). Some much smaller holes, 0.5–3.0 µm are also thought to be caused by amoebae and these overlap in size with holes caused by some bacteria. Amoebae have been shown to attack many fungi including the pathogens *Fusarium* spp., *Botrytis* spp., *Cochliobolus sativus* and *C. miyabeanus, Thielaviopsis basicola, Phytophthora cinnamomi, Gaeumannomyces graminis, Rhizoctonia solani*, and *Verticillium* spp. Exactly how they make holes is not clear, there must be a concentration of wall degrading enzymes in

Fig. 5.15. Holes in the hyphae of the pathogen *G. graminis* caused by mycophagous amoebae. (Photograph courtesy of Cook, R. J. From Homma, Y. *et al.* (1979). *Phytopathology* **69**, 1118–22.)

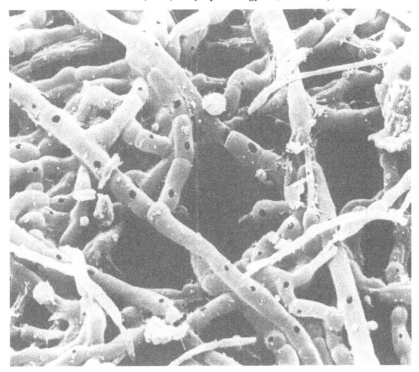

the cell surface membrane. The holes are not dissolved out, they are cut to remove a disc of wall material. The amoeba stops over a hyphal compartment or spore and for a while nothing appears to happen, but if the amoeba is disturbed a circular depression is found in the wall. Eventually this circular trench cuts right through and a disc of wall material suddenly appears in the amoeba. Later the amoeba leaves the disc behind as it moves away. The amoeba enters the cell through the hole in the wall and cleans out the cytoplasm. The whole process takes a few minutes.

The amoebae must have water-filled pores to live and move in, so the soil needs to be quite wet. Just how wet is in some dispute, and no doubt it depends partly on the species of amoeba and hence the size of pore it needs (Fig. 5.16), but there are some active down to about −100 kPa. Most have optima much wetter than this, perhaps no more than −20 kPa. Depending on the clay content of the soil, this can mean that they are only active in soils wetter than field capacity, though as cysts they can survive almost complete dryness. They could be useful for diseases (e.g. caused by oomycetes with zoospores) that flourish in wet conditions.

Though they are found in most soils worldwide, their numbers may be quite low, usually no more than a few hundred per gram of soil.

Fig. 5.16. Activity of amoebae in a sandy loam (○) and a heavy clay loam (●) at various matrix water potentials. Unbroken line (————) represents total amoebae and broken line (– – – –) represents mycophagous amoebae. (From Old, K. M. & Chakraborty, S. (1986). *Progress in Protistology* **1**, 163–94.)

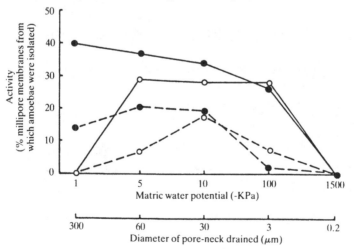

Various workers have shown reductions in the populations of pathogen and correlations between the number of amoebae and the suppressiveness of soil, but there is no very convincing experimental evidence that the addition of amoebae to soil reduces the amount of disease to a significant degree.

Even if amoebae could be shown to be effective there is some doubt about whether they could be produced commercially. Their production would not be so easy as bacteria in normal, stirred, bulk fermenters, but it may be possible to use some of the technology now developed for the culture of animal cells used in the production of monoclonal antibodies. The amoebae would presumably be prepared as cysts which would be very tolerant of adverse conditions and have a long shelf-life.

Finally there is the problem of how selectively the amoebae feed. Might they not consume useful organisms as well as pathogens? There is evidence that they reduce the effectiveness of *Pseudomonas* inoculated as a plant growth promoter. This possible destruction of useful organisms is a potential problem with most biocontrol agents, especially parasites and predators.

So, despite the intrinsic appeal which these protozoa have, there is really little hope for using them commercially in the foreseeable future. There may be some specialist applications in horticulture, and perhaps paddy rice might be worth examination, but it seems more likely at the moment that one of the bacterial or fungal antagonists will be used rather than amoebae.

5.8   Plant growth promoting rhizobacteria (PGPR)

There have been reports for many years of increases in growth and yield of plants after inoculation of bacteria. Initially these were linked to nitrogen fixation by organisms such as *Azotobacter*, but this is now largely discounted mostly on the grounds of carbon limitation for this process with a very high energy demand. Bacteria were also used that were said to promote phosphorus solubilization, but it seems unlikely that they have any enzymes that the plants do not already possess. Many bacteria that are isolated from soil can produce plant growth hormones in culture, and it is possible that these can affect root growth.

There is now no doubt that there are bacteria that can be isolated from soil which, when back-inoculated, stimulate plant growth. Some of the initial results were very spectacular (Table 5.9) with increases in the grain yield of oats and wheat and very great increases in the yield of carrots which were moreover of larger sizes and therefore more profitable to market. In glasshouse trials on potatoes there were reports of 500% yield increases though in field trials this went down to less than 20%. Still 20% yield increases in one operation are far above even the wildest dreams of any plant breeder. There was great variability, and it was noticed that these increases did not occur in peat or in sterilized soils. Table 5.10 shows that only about half the trials produced an increase in growth. In some of the worse series only one trial in nine was successful. This variability is the same problem that has occurred in all attempts to introduce bacteria into crop systems where they are not

Table 5.9(*a*) *Effect of inoculating seed with* Streptomyces griseus *or* Bacillus subtilis *on grain yield of oats and wheat*

	Grain yield, kg	
Seed treatment	Oats	Wheat
S. griseus	23.2	108.4
B. subtilis	22.4	103.4
Water	16.0	100.6
Dry	16.9	101.3
$P = 0.05$	2.8	21.7
$P = 0.01$	3.8	29.3

Table 5.9(*b*) *Effect of the same organisms as in (a) on the yield size of carrots*

		Yield, t/ha[a]			
Seed treatment	Total	Very large	Large	Medium	Small
B. subtilis pellet	83.75	18.00	29.25	21.88	14.63
S. griseus pellet	53.25	2.75	9.63	17.00	24.03
B. subtilis	63.38	7.63	17.13	21.25	17.38
S. griseus	62.75	6.88	20.63	19.88	15.38
Pellet	56.75	2.88	11.75	20.00	22.13
Water	53.75	5.00	13.38	18.00	17.38
$P = 0.05$	7.83	3.73	5.15	5.50	5.23
$P = 0.01$	10.53	5.00	5.90	7.38	7.00

[a] Very large, 5.72 × 20.32 cm; large, 4.45 × 20.32 cm; medium, 3.18 × 15.24 cm; small, 1.91 × 13.70 cm.

From Merriman, P. R. *et al.* (1975). In *Biology and control of soil-borne plant pathogens*, ed. G. W. Bruehl, pp. 130–3. St. Paul, Minnesota: American Phytopathological Society.

present (see section 5.1). A particular strain would work in one year at different places in California and Idaho, but the next year nothing happened anywhere with that strain. Some work consistently from year to year, but only on one site or soil type.

Most of the organisms used in these trials were fluorescent *Pseudomonas* spp., especially *fluorescens* and *putida* and there was evidence that siderophores were again involved, though the main one, pseudobactin, also has antibiotic properties and the relative roles of antibiosis and chelation may not be determined. They have been shown to colonize the root systems of treated plants, though not always very evenly, and to persist throughout the growing season in numbers around 10^4 cfu cm^{-1} root which is much above the natural level of pseudomonads.

The main mechanism is now thought to be by the control of 'minor pathogens' and deleterious bacteria. There are fungi and bacteria in soil that may cause a reduction in growth or some slightly debilitating condition without being obvious enough to have been recognized and described as a disease. It is this rather nebulous group of organisms that are thought to be discouraged or reduced by the PGPR, so allowing the

Table 5.10 *Comparison of the overall success ratio of various strains of PGPR in field trials of potato and sugar beet*

Strain	Crop	Significant[a] trials/total	Average[b] increase (%)
TL3	Potato	6/11	17
BK1	Potato	2/8	14
TL10	Potato	1/9	33
A–1	Potato	4/5	9.5
	Sugar beet	2/4	11.0
B10	Potato	2/4	12.5
E6	Potato	2/3	10.0
	Sugar beet	1/2	6.0
SH5	Sugar beet	5/9	11.6
RV3	Sugar beet	4/7	10.2
B4	Sugar beet	3/7	11.0

[a] Number of trials in which each strain was tested where significant increases in final yield were attained.
[b] Mean percent increase in yield compared to untreated or fungicide treated controls for trials in which significant differences were attained.
Compiled from various sources by Suslow, T. V. (1982). In *Phytopathogenic prokaryotes*, vol. 1, ed. M. S. Mount & G. H. Lacey, pp. 187–224. New York: Academic Press.

plant to express its full growth potential. As such the PGPR are biocontrol agents, controlling these minor pathogens and deleterious organisms and are not actually having any direct effect in promoting plant growth.

If PGPR are a reality there is also a way in which they can help in the control of recognized root disease. If the roots are being killed by a pathogen some plants, such as cereals, may produce more roots. If their growth can be promoted to the extent that they produce roots faster than the pathogen kills them, then the plant will live. The disease has not been cured but the plant can outgrow the disease that is there.

Another facet of the expanding PGPR story is that there are seedling emergence promoting bacteria which, especially under cold damp conditions, increase the germination and early growth of seeds sown at the limit of their climatic range. It is possible that these are controlling incipient damping-off or similar problems.

The most carefully researched case of PGPR is that of potatoes in Holland (Schippers *et al.*, 1987). If this crop is planted repeatedly on the same field or in short rotations the tuber yield decreases by at least 10%. This is not to do with nutrient deficiencies or simple disease effects. Inoculation with several different strains of *Pseudomonas* corrected much of the yield loss, but only in these soils which are overcropped to potatoes (Table 5.11). There is no effect in soils that have had long crop rotations, so it is not growth promotion so much as the correction of a growth deficit. The *Pseudomonas* strains used were not known to produce much, if any, antibiotic and the main activity was due to siderophores. This is based in part on the usual tests with other chelators (see section 5.6.3) and on the use of Tn5 mutants, which did not produce siderophores and which were no longer effective. Some care is needed here, however, for it is known that even quite specific changes brought about by these genetic engineering techniques may do several things at once, i.e. transfer more than one gene or gene sequence. Thus in *P. syringae* apparently single gene Tn5 insertions transferred both siderophore production and the ability to produce the antibiotic syringomycin. However, given that the pseudomonads that cure the potato problem seem to rely mainly on siderophores, how do they work? There are bacteria, often also pseudomonads, in most soils that can produce cyanide, but these deleterious pseudomonads are not usually fluorescent. They are particularly common in the soils cropped repeatedly to potatoes, where about 50% of *Pseudomonas* isolates may be deleterious, though cyanide production has not been detected in these soils with the methods of analysis available. Cyanide production

requires iron, so the theory is that the competition with the PGPR
reduces the cyanide producers or the amount of cyanide that a given
population can produce. This is summarized in Fig. 5.17 which shows
how the competition between the two sorts of bacteria results in an
improvement of the plant growth when this was previously inhibited by
proposed cyanide toxicity in the roots.

These deleterious bacteria perhaps deserve a little more attention.
Many genera have been implicated, not just the non-fluorescent
pseudomonads described above (e.g. *Bacillus, Streptomyces, Enterobacter, Klebsiella, Arthrobacter, Flavobacterium*, etc.) and they may be
very common in soils, forming a considerable proportion of the
population. There are reports of up to 48% reduction in shoot growth,
reduced root growth, reduced germination, root distortion and more
infection by other pathogens. In extreme cases there could be chlorosis
and wilting of the leaves and epinasty similar to the effects of ethene
(ethylene). Many different plant species can be affected, but there also
seems to be some effects that are cultivar specific or at least some
cultivars are more seriously affected than others by the deleterious
bacteria.

Table 5.11 *Effect of tuber treatment with antagonistic fluorescent*
Pseudomonas *isolates (WCS) on plant growth and yield in soil from a
field continuously cropped with potato and therefore containing the
inhibitory organisms. Control was treated with carboxymethyl cellulose
(CMC) adhesive only*

Seed tuber treatment	Shoot dry weight %	Root dry weight %	Tuber fresh weight[a] %	Total dry weight[a] %	Number of tubers[a] %
CMC (control)	100	100	100	100	100
WCS 307	102	108	241	112	320
WCS 377	103	101	366	116	900**
WCS 361	89	103	332	107	666**
WCS 358	113	116	338	128*	834**
WCS 374	100	101	469**	123**	934**
WCS 365	102	121	550**	131***	866**

* Significant for $P = 0.05$; ** $P = 0.025$; *** $P = 0.005$. Significance was
tested by means of analysis of variance.
[a] Control weights and number of tubers >5 mm were respectively: 7.8 g, 1.8 g,
4.3 g, 10.2 g and 0.5.
From Geels, F. P. & Schippers, B. (1983). *Phytopathologische Zeitschrift* **108**,
207–14.

This whole subject of PGPR has had several false, or at least dubious, starts and seems anyway to be not so much growth promotion as removal of growth inhibition. The suggestion that siderophores are involved still has many supporters and there is good evidence for them in some cases, but there is a danger that one mechanism is found and everyone then assumes that is the most important. There may well be other mechanisms just as important, or more so, that have yet to be well demonstrated. The association of PGPR with deleterious bacteria may explain some of the great variability: they only work when there is something to work against, as in the over-cropped potato soils. We may be back to trying to decide whether it is the organism that has failed the test, or whether the test has failed to set conditions in which the organisms can be expected to work (section 5.6). We end again on the note that there definitely seems to be something there. If the yield increases are only half as good in commercial use they are worth trying for, but they must be consistent to be of any value.

Fig. 5.17. Diagram of the hypothesized interactions between plant growth promoting (PGPR) pseudomonads and the potato root in short potato rotations (potatoes grown frequently or continuously). Harmful rhizosphere micro-organisms (HMO), probably pseodomonads, can produce cyanide (CN^-) which inhibits the cell energy metabolism and thereby plant growth. Siderophores (Sid) released by the PGPR decrease the availability of the Fe^{3+} thereby inhibiting CN^- production by the HMO and increasing plant growth. (From Schippers, B. (1986). In *Microbial communities in soil*, ed. V. Jensen *et al.* pp. 35–48. London: Elsevier Applied Science Publishers.)

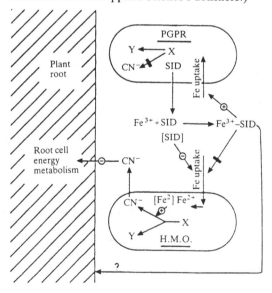

5.9 Integrated control of root diseases and interactions amongst pathogens

Integrated control is the combined use of chemicals and biocontrol agents to control disease but it is little used or reported for roots (though see for glasshouse diseases, section 7.4). Some of the chemical treatments recommended to control disease do in fact work via micro-organisms and these are the oldest examples of this method, even if the biological involvement was not understood at the time.

Armillaria mellea, the honey fungus, causes serious root rots of many woody species and is very difficult to treat and eradicate. Soil fumigation has been used on valuable crops, especially fruit trees. As long ago as 1951 it was noticed that when the soil in citrus orchards was fumigated with carbon disulphide the roots with dead *Armillaria* in them almost invariably had *Trichoderma* as well. It was deduced that the *Trichoderma* survived the fumigation, as did some of the rhizomorph tissue of honey fungus. Later the *Trichoderma* proliferated, free of competition from other soil organisms, and killed the honey fungus. A better documented case of the same process was observed when methyl bromide was used for the fumigation. The concentration of fumigant can be much lower than is needed to kill the *Armillaria* as the stimulation of *Trichoderma* occurs even in only partially sterile soil (Fig. 5.18) and then acts as a parasite on the weakened pathogen. There is now a range of isolates or mutants of *Trichoderma* which are resistant to methyl bromide, pentachloronitrobenzene (PCNB), benomyl or captan. These can be inoculated at the time of a fungicide treatment to reinforce the chemical control, and usually this means that lower levels of chemicals, or less frequent application, can be used.

There are a number of fungi that can attack cereal stem bases and leaves all of which are soil- or trash-borne organisms which interact with each other and with fungicides. This is not so much integrated control as iatrogenic disease (disease caused by man's activities). There is take-all (*Gaeumannomyces graminis*), *Fusarium* spp. especially *F. culmorum* causing foot rot, *Rhizoctonia cerealis* (formerly considered a race of *R. solani*) causing sharp eyespot, and *Pseudocercosporella herpotrichoides* causing eyespot. Normally take-all is a major problem. Sharp eyespot, eyespot and foot rot usually cause little loss in vigour when they occur alone, but each may make the take-all loss worse. The pathogens may occur individually on a plant, but about half of the plants with eyespot also have *Fusarium*. It is known that in culture the *Rhizoctonia* is inhibited by the others and so it normally occurs alone on a plant. Eyespot may be controlled by systemic fungicides, but extensive resistance has developed in the field. Reduction of eyespot by fungicide leads to worse attacks of foot rot and sharp eyespot, which were previously insignificant. Similarly fungicides such as benodamil which decrease *Rhizoctonia*, make the other two diseases much worse. So the level of one pathogen affects the amount of colonization of others, and selective fungicides, or anything else that affects the level of a particular member of the disease complex, will alter the balance between the pathogens. Added to this is *Cephalosporium*, which occupies stem bases and prevents entry of other pathogens (section 1.2), including those just described. Apart from these pathogen interactions there is also the possibility that fungicides are removing natural saprotrophic antagonists. One potential antagonist is the fungus *Microdochium bolleyi* that has been shown in experiments to control eyespot and is also a potentially useful antagonist for take-all. It may be better to try to control these stem base diseases by the manipulation of the pathogens

Fig. 5.18. Percent of isolations giving *Armillaria* or *Trichoderma* from roots infected with *A. mellea* kept in non-sterile soil after treatment with methyl bromide. (From Ohr, H. D. *et al.* (1973). *Phytopathology* **63**, 965–73.)

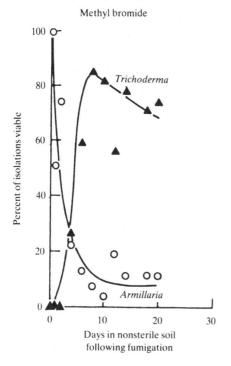

themselves or saprotrophs in the soil, since fungicides have given resistance problems and made some diseases worse. This whole complex of diseases is certainly worth further study.

5.10   Conclusions

I have just summarized a small selection of the work which has been conducted on the biocontrol of soil-borne and root diseases of plants. It is noticeable that there are very few commercially available products, and those that do exist are for specialist markets in horticulture. There are many possible reasons for this which have been discussed before (section 5.1) but it should also be remembered that chemical control of soil diseases is not so successful: there are some diseases for which there are still no control measures and others where, compared with leaf diseases, the approach is somewhat crude (soil fumigation with rather nasty, unselective toxins). Indeed this is the very reason for much of the research into biological methods, where there is no competition from chemicals for the market.

What is the future for biological control in soil? The first attempts at commercial use for major diseases are imminent. As usual the problems have become much more complex than they at first appeared and we are only now beginning to understand, with the increased knowledge of microbial ecology in general, how delicate is the balance between micro-organisms in soil and how many climatic and edaphic factors affect it. Some say that we are wasting time and resources in an empirical approach (go out and isolate some organisms and give them a try) and that we should wait for a more complete understanding of the environment and microbial interactions before we seriously try to find biocontrol agents. That would probably mean waiting a very long time. Anyway chemicals have traditionally been found on an almost random selection and testing basis, and no-one can deny that there are many chemicals that have successfully controlled plant diseases and given great increases in world food supply. Industry, which will eventually fund the product development and marketing, is used to the screening and testing of potential products, and the resulting tests for the environmental protection agencies before clearance. Perhaps biocontrol should adopt the same methods that have led to good results for the control of many diseases up to now. The alternative is a new type of agricultural and agrochemical industry based on sustainable production without the intense use of fertilizers and pesticides, which would need, and could lead to the development of, different biocontrol strategies.

This has been a quite long chapter (even without many commercial products to discuss!), but it does reflect the amount of work in this area. Rightly or wrongly many researchers, and funding agencies and industry, see biological control being of use in soil, so the research will continue and will, in the long term, be successful.

6

Biocontrol of diseases of flowers and fruits

6.1   Introduction

Flowers are ephemeral structures and do not themselves suffer from many diseases, though they are points of entry for pathogens. Fruits and seeds on the other hand are a major world food which may be on the plant for a long time and may then be stored. They are subject to a large number of diseases, both during growth and post harvest and this, together with insect attacks, represents an enormous loss (about one-third of production, even with the use of pesticides) which is serious in both economic terms and in the human and animal suffering caused by starvation. If even post harvest losses could be reduced or eliminated the world's food problems would be solved. We can, and do, produce enough food but without expensive storage facilities much is lost, especially in the tropics with ideal conditions for decay but often the inability to pay for chilled stores and other expensive means of storage.

There are a large number of chemicals potentially capable of controlling the spoilage organisms, which are mostly fungi, but there are serious toxicity problems. Fruits are grown to be eaten and, especially with post harvest rots, the problem may be at its most serious close to the point of consumption. There are, therefore, limitations on the sort of chemicals that can be used. There is less of a problem in the field because of the longer time between application and consumption. The same care needs to be taken with biological control agents because application of large numbers of fungal spores or bacteria to protect fruits could lead to the ingestion of their metabolic products or the organisms themselves. Thus there are reports of control of storage rots of citrus by the application of *Bacillus subtilis* in washing water, but this has not become commercial. In general, naturally occurring organisms should be used, with a short life expectancy and as long a time separation as possible between application and consumption.

6.2   Flowers

Flowers can be infected by wind or insect transmitted diseases and they may pass the organisms on to the fruit or seeds which are formed. Most fruits and seeds have an internal bacterial and fungal flora which are harmless saprophytes derived from the flower. There is, however, one serious disease that arises primarily from flower infection and that is fire blight of rosaceous plants which has been mentioned previously. The causal organism, *Erwinia amylovora*, occurs also on leaves and may cause stem cankers (sections 3.7 and 4.1). It is most serious on pears, though it does occur on many other species and is much affected by the environmental conditions at the time of flowering.

The bacterium overwinters in stem cankers and in latent infections in stems and is then transferred in spring to the opening blossoms by insects and, to a lesser extent rain splash. Once in the flowers *E. amylovora* multiplies in the nectaries to form a secondary source of inoculum which is transferred from flower to flower by bees and other nectar feeding insects. The infected flowers die and infection also enters the pedicel and eventually the stem to form the cankers. The obvious control measure is very good sanitation with the removal of the cankers containing the primary inoculum, though these may be very difficult to see. Cutters used in pruning should also be sterilized by dipping in disinfectant. There always remains, however, a small reservoir of infection which is sufficient to start an epidemic. The disease has been known in America for 100 years and was introduced into Europe in 1957: sanitation failed to contain the outbreak and the disease is now endemic. Chemical control is difficult, for there are few agriculturally available bactericides, except various copper compounds. In the United States streptomycin is used but there are environmental objections to the widespread dispersal of antibiotics that may increase the number of bacteria resistant to a still medically important therapeutic agent. Streptomycin is also expensive.

Biological control has been used successfully, the agent usually being the closely related *E. herbicola* though *Pseudomonas* has also been used combined with the prevention of ice damage (see section 3.7). This is one of the longest running programmes for a potential control agent, the first attempts being made more than 50 years ago (Beer, Rundle & Norelli, 1984; Lindow, 1985). In general, the *E. herbicola* is sprayed onto the flowers at, or preferably just before, the time of potential infection and it occupies the same niche as the pathogen, apparently reducing the numbers of *E. amylovora* by exploitation competition. There is also evidence for the production of bacteriocins by some strains of *Erwinia*. Other less well defined inhibitors are also present in some culture filtrates (Table 6.1). The control can be good, though it may be overwhelmed by very high densities of the pathogen. The reduction in the pH of the nectar, caused by the growth of the antagonist, may also be important in limiting the pathogen. The control of flower infections that can be achieved compares favourably with commercial bactericides, though the volume of bacteria sprayed was very high and repeat applications were necessary (Table 6.2).

The reduction in *E. amylovora* on leaves is also important in reducing secondary infections and the overwintering population and this is achieved by *Pseudomonas syringae* and various other bacteria (Fig. 3.14). In this case the effect was not dependent on the production of antibiotics. Fruits may also be infected from the flowers and these have been used in a screening system for antagonists to *E. amylovora* (Beer *et al.*, 1984).

There needs to be some further development of this system but it is very likely that there will soon be commercially available a combined system for fire blight and frost damage control, possibly involving more than one organism.

Table 6.1 *The effects of live cells and the culture filtrates of three isolates of bacteria (yellow rods, numbers c/1, c/81, 112y) tested against fire blight (*E. amylovora*) which was inoculated at various densities*

| | Inoculum density (cells/ml) of the pathogen | | | |
	10^4	10^5	10^6	10^7
	Percent blighted shoots			
Untreated control	85	100	100	100
c/1 live bacteria	5***	25***	40***	90
filtrate	50*	64*	95	100
c/81 live bacteria	0***	15***	40***	75*
filtrate	50*	75*	98	100
E. herbicola 112 y live bacteria	5***	35***	55***	85
filtrate	35**	50***	89	100

* ** *** = significantly different from untreated control at 5%, 1% and 0.1% probability, respectively.
From Isenbeck, M. & Schulz, F. A. (1986). *Journal of Phytopathology* **116**, 308–14.

6.3   Fruit

There are general rots of fruits by saprotrophs or opportunistic pathogens such as *Botrytis, Rhizopus and Penicillium* (*r*-strategists), and also more specialist pathogens like *Monilinia* on rosaceous fruits, and coffee berry disease caused by *Colletotrichum coffeanum.*

The latter is a good example of biological control detected by the adverse effects of fungicides on natural antagonism and it has been mentioned previously (section 3.3). Sprays for coffee rust make berry disease worse and necessitate further sprays, because of a reduction in antagonists on bark which normally limit the sporulation of the pathogen. There also seems to be a more complex interaction in which early infection of the berry normally leads to non-sporulating lesions that give some induced resistance to the fruit and prevent the more serious, later infections which produce the spore inoculum for survival.

Monilinia causes a brown rot of apples, pears, plums and peaches which usually have a very characteristic appearance with concentric

Table 6.2 *The effect of control measures, including saprotrophic bacteria, antibiotics and commercial test compounds, on reducing the colonization of pear flowers by* E. amylovora *and in reducing the incidence of fire blight in Bartlett pear trees*

Treatment	Amount applied (per hectare)	Flowers colonized with E. amylovora[a,b] (%)	Number of bacteria per colonized flower ($\times 10^6$)[b]	Flower infections (no. per tree)[b]
Control	...	52 wx	1.5 w	3.2 w
Citcop	9.28 litres	51 wx	1.3 w	1.7 xy
Citcop	18.56 litres	48 wx	1.2 w	2.2 wx
Saprophytic bacteria[c]	464.00 litres	46 wx	0.8 w	1.0 xyz
Terramycin (17%)	0.70 kg	40 wx	0.9 w	0.8 yz
Streptomycin (17%)	0.70 kg	37 wx	1.5 w	0.8 yz
Kocide	0.56 kg	32 wx	0.8 w	1.0 xyz
Kocide	2.24 kg	28 x	0.4 w	0.4 z
MBR 10995 (25%)	0.28 kg	19 y	0.4 w	0.4 z
MBR 10995	1.12 kg	12 y	0.2 w	0.4 z

[a] Correlation coefficient (*r*) between the percent of flowers colonized and number of infections per tree is 0.87, significant $P = 0.01$.
[b] Values followed by different letters are significantly different, $P = 0.05$, as determined by Duncan's multiple range test.
[c] Approximately 3.2×10^7 cells per millilitre of suspension of each of three saprophytic pseudomonads and a saprophytic *Erwinia* sp. were applied every 5 days from 16 March through 18 April for a total of eight applications.
From Thomson, S. V. *et al.* (1976). *Phytopathology* **66**, 1457–9.

rings of spores over the soft brown fruit surface (Byrde & Willetts, 1977). It is a most serious disease in orchards where infection takes place in the spring with conidia and possibly ascopores, produced from the mummified fruits left on the ground. Young shoots and especially the blossom are attacked and later infection is passed to the developing fruit where it may become latent until the conditions in storage favour spread. If the rot develops directly, the fruit falls off the tree prematurely and becomes a brown, hard mummy composed mostly of fungal material in the remains of the fruit which survives over the winter. *Trichoderma viride* has been shown to control *Monilinia* and various *Bacillus* species are known to be antagonistic to the brown rot fungi, especially by producing antibiotics and reducing the longevity and germination of the spores of the pathogen. The bacteria themselves have been used and also the antibiotics from culture filtrates, but there has been no commercial development, probably because fungicides are routinely used in orchards for other diseases and they give some control of *Monilinia*.

The most serious problem with the production of soft fruits is a wide variety of post harvest rots which gives the product a very short shelf-life (Dennis, 1983). By far the most important organism is the common grey mould *Botrytis cinerea* which also causes field infections. There may also be infections with *Mucor* and *Rhizopus* especially late in the season and when fungicides such as benomyl, dichlorofluanid or dicarboximides have been used on *Botrytis* (they are not toxic to phycomycetes). *Botrytis* itself develops resistance to these fungicides in the field. There is, therefore, a serious disease complex where the initial control by fungicides may break down and alter the balance of importance between the different pathogens involved.

Botrytis has been most studied on strawberry, where infection of the flowers or the fruit usually occurs from saprotrophic growth on crop debris. Infections may become latent on the withered flower parts, only being expressed in favourable conditions when the fruit is ripe and the rot starts at the calyx. *Botrytis* can grow at low temperatures and may be a problem even in chilled storage. Various control agents have been used including *Trichoderma* spp. which gave as good control as standard fungicides (Table 6.3). Different antagonists were used in a study of the biocontrol of *Botrytis* on tomato, where *Cladosporium herbarum* and *Penicillium* gave good results (Table 6.4). *Enterobacter cloacae* reduces *Rhizopus* rot in fruit by 50%, but there are doubts about its use on uncooked food, even though it is a normal inhabitant of fruit and is a common gut organism.

In connection with fruit rots there is an interesting study by Swinburne (1986) and co-workers that throws a slightly different light on some mechanisms proposed for biological control of plant pathogens. Banana fruits are rotted by a variety of fungi, but especially *Colletotrichum musae* which infects through the skin of the banana. Its germination is stimulated by leachates from the fruit, especially anthranilic acid, which may act as a siderophore, sequestering iron. Germination and appressorium formation are also stimulated by the siderophores from *Pseudomonas fluorescens* and other bacteria on the fruit surface. Free iron, and various chelates in which iron is available to the plant, inhibit germination, possibly by stimulating phytoalexin production. Here we have the exact reverse of the mode of action proposed for some biocontrol agents whose siderophores reduce infection by starving the pathogen of iron (section 1.3.2). If biological control was ever used for this rot of banana fruit, the biocontrol agent would have to be specially selected to be *without* siderophores and would work by out-competing a natural flora that did produce these germination-promoting chelating agents. One should be wary about generalizations on biocontrol agents!

Apart from storage rots *per se* there are infections of fruits and seeds that may be serious because of the toxic metabolic products that are left behind. In particular a lot of attention has been given to mycotoxins, especially aflatoxins which are named from the toxin produced by

Table 6.3 *The effect of different isolates of* Trichoderma *spp. on rotting* strawberries by Botrytis *and* Mucor mucedo, *compared with standard fungicide treatment*

| | Percent of rotten fruit | | |
	Botrytis cinerea	*Mucor mucedo*	Total
Untreated	19	2	21
Dichlorofluanid	11	3	13
Trichoderma pseudokoningii 13	15	3	18
T. hamatum 85	12	2	14
T. harzianum 107	11	2	12
T. viride 1	12	1	12
T. viride 2611	14	2	16
		LSD for total 4.5%	

From Tronsmo, A. & Dennis, C. (1977). *Netherlands Journal of Plant Pathology* **83** (Suppl. 1), 449–55.

Aspergillus flavus, though many other fungi are now known to produce secondary metabolites with harmful effects on animals and man who eat the affected seed. Chemically they are a diverse group of compounds (Purchase, 1974). The growth of the fungi can be controlled by sanitation and by the storage conditions post harvest. The initial infection by deleterious fungi is normally acquired in the field or during harvest and the natural microflora including yeasts, *Ulocladium, Fusarium* and other species of *Aspergillus* can limit the numbers of *A. flavus* on almonds which are later stored. There is again no commercial exploitation of this phenomenon, because there are already good control measures by manipulating storage conditions. There must also be some doubt whether the control might be as bad as the disease for some strains of *Fusarium* and *Ulocladium* are themselves known to produce toxins, so careful testing would be necessary. We are back to the original problem of applying control agents, or chemicals, to a food that is about to be eaten.

This chapter has, necessarily, been short because there is little research on the biological control of diseases of flowers and fruit, and even less practical applications. The exception to this, because of a lack of good chemical methods, is fire blight where present work looks promising. The future for the other diseases depends largely on the developments, or otherwise, in chemical control. At present there are

Table 6.4 *The effect of* Cladosporium *and* Penicillium *on the rot of tomatoes caused by* Botrytis

	Total number of fruit	Number rotted	Number infected on petals only	Number of fruit with no infection	% infection
Trial 1					
Check	124	39	18	67	46
Cladosporium					
herbarum	85	0	1	84	1
Penicillium	81	0	1	80	1
Trial 2					
Check	83	55	11	17	80
Cladosporium					
herbarum	90	3	0	87	3
Penicillium	68	2	0	66	3

From Newhook, F. J. (1957). *New Zealand Journal of Science and Technology* Ser. A **38**, 473–81.

many fungicides used very frequently in commercial orchards, and as long as this is permitted by law, or is required because the customer demands a 'perfect' fruit free of all blemishes, then there will be at least some control of diseases like *Monilinia*. The situation may change if the permitted levels of pesticide residues are reduced or if the customer will accept minor blemishes, such as some scab on apples, which does no harm and does not cause rotting. There may be similar problems in orchard crops as have developed in soft fruits where the resistance of *Botrytis* to a number of chemical fungicides is a continuing problem. Whatever the reason it seems that there are several potential biocontrol agents which could be used if required, though much further development would be necessary. Control would have to be in the field, during the growing season or by the reduction of overwintering inoculum so that the possible problems with control on the fruit near the point of consumption are avoided.

7

Biocontrol of diseases of seeds and seedlings

7.1   Introduction

The main diseases that we are concerned with in this chapter are seed rotting, pre- and post-emergent damping off and various seedling blights. This complex of diseases is mostly caused by a few genera of fungi, especially *Rhizoctonia* and *Pythium*, with *Phytophthora*, *Fusarium* and *Sclerotinia* causing less widespread problems. These are unspecialized pathogens (*r*-strategists) which use exudates from the germinating seeds for saprotrophic growth before they attack the very young plants that have not developed effective mechanical barriers to infection. *Rhizoctonia* usually attacks the seed, hypocotyl or stem, while *Pythium* attacks the root tips. Under very wet conditions *Rhizoctonia* and *Phytophthora* may grow amongst the tops of the seedlings. The diseases are especially bad if conditions are not favourable to the rapid growth of the seedling. Damping off is characteristic of crops that have been sown too early in the year so that they are germinating slowly in damp soils at low temperatures. Vigorous seedlings getting away to a good start under ideal conditions do not usually suffer from these problems.

However, often there are agricultural reasons for trying to make an early start to the growing season, or seeds sown in a warm period of weather may be overtaken by a cold wet spell. In glasshouses, where many seedlings are raised for horticultural crops, there are less climatic problems, but heating costs are high and some growers may try to manage with just a little less heat than is best, or seeds may be overwatered. Seed rots and damping-off remain a real problem.

There are cheap, effective chemical means of control, by applying fungicides to the seed coat as a dust, with various adhesives or possibly with mineral nutrients as well in pelleted seeds. In horticulture there is also the possibility of using composts or soils that have been artificially 'sterilized' by steam or chemicals. They are very rarely strictly sterile, but they may contain many fewer propagules of the pathogens than normal. The problem with such growing media is that they are a biological vacuum, and are subject to very rapid colonization by some micro-organisms (*r*-strategists) which is good if these are *Trichoderma* or *Pseudomonas*, but they may also be *Rhizoctonia*, *Fusarium* or *Pythium* leading to a worse disease problem than before the sterilization. *Pythium* especially may survive some of the steaming treatments. The answer to this has been to only partially kill the microflora, in the hope of killing the pathogens and leaving the useful organisms. This may be done by injecting steam mixed with air into the soil to raise the temperature to about 60°C; pure steam is too hot and also kills the saprotrophs. In countries that have reasonable amounts of sunshine this may also be done by covering the soil with tarpaulins or black polythene to raise the temperature of the surface layers (called solarization).

Any biological control system has to work in competition or co-operation with these established methods of control. For example, the soil may be chemically or steam treated and then a potential control agent, or at least a harmless saprotroph, is introduced to ensure that recolonization is not by the pathogen. There are, however, many purely biological control measures that have been developed for these diseases and examples of biological control of *Rhizoctonia* and *Pythium* in the glasshouse are common in the literature, but field control by biological means is less often reported. Seeds and seedlings in *in vivo* screening systems are easy and quick to work with in the laboratory and glasshouse, so programmes for developing control agents can be planned and budgeted to fit in with the funding bodies accounting procedures. Fortunately there are also less cynical reasons why biological control of seed and seedling diseases is a good proposition.

Antagonists are quite easy to apply to the seed with existing seed-coating technology and they get straight to the place of action: there is no problem with getting them suitably positioned in adequate numbers as there is with root diseases especially (Chapter 5). Secondly, the protection is only needed for a short time in most cases, perhaps as little as one or two weeks until the seedling is mature enough to have defences against these opportunistic pathogens: so survival or multiplication of the antagonist is not a problem. There is the possibility, mentioned above, of modifying the environment to favour the antagonist, especially in horticultural conditions. Finally there are reports of soils that are naturally suppressive to damping off and seedling blights, so there is a good starting point for the search for potential control agents in these cases where control is occurring naturally (see below).

The antagonists that have received the overwhelming amount of attention are in the genus *Trichoderma*, but *Bacillus*, *Streptomyces* and *Gliocladium* have also been used. These are all fast growing saprotrophic *r*-strategists that may successfully compete with the pathogens for the exudates from the seeds, and some are antibiotic producers and mycoparasites. The site of activity seems to be the actual plant surface where the infection is taking place, and control is not usually by the reduction in inoculum levels. Damping off, especially that caused by *Rhizoctonia*, is one of the few diseases where *Pseudomonas* is not considered a good antagonist, though it can be effective against *Pythium*. The antagonists may be strain specific, or rather some strains of *Rhizoctonia* are unaffected by otherwise effective control agents.

7.2   Suppressive soils

The phenomenon of suppressive soils, and the opposite condition of conducive soils, has been much written about and discussed (Gerlagh, 1968; Schneider, 1982; Cook & Baker, 1983; see Chapter 5). A soil suppressive for one pathogen, or even one race of a pathogen such as *Rhizoctonia*, may not be equally suppressive for other diseases or other races. This is specific suppression.

Both *Pythium* and *Rhizoctonia* need saprotrophic growth before infection and organic substrates favour this and so increase disease. There is, however, competition with saprotrophs exploiting the same substrates and this may lead to the pathogens being suppressed. Suppressiveness can often be increased by repeat planting of the same crop so that the disease is at first worse, then gradually gets better (Fig. 7.1). The time-scale is very variable: in some cases with field crops it may be years (e.g. with take-all, section 5.2.2) but with radishes it may appear in only weeks after repeated replanting. Soils naturally suppressive to *Rhizoctonia* causing damping off of radish, alfalfa and sugar beet have been found to have high levels of some species of *Trichoderma* and the degree of suppressiveness can be directly related to the amount of the supposed antagonist. The obvious step is to introduce the *Trichoderma* during the course of replanting and so speed up the development of control (Fig. 7.1) and even conducive soils can be made suppressive by the addition of *T. hamatum* to control *Rhizoctonia*, *Pythium* on peas and *Sclerotinia* on beans.

Fig. 7.1. Damping-off of radishes when soil (Fort Collins clay loam) conducive to *Rhizoctonia solani* was infested with conidia of *Trichoderma harzianum* (Israel isolate) and replanted weekly for 8 weeks. At the end of the period the levels of soil suppressiveness were significantly different, $P = 0.05$. (From Liu, S & Baker, R. (1980). *Phytopathology* **70**, 404–12.)

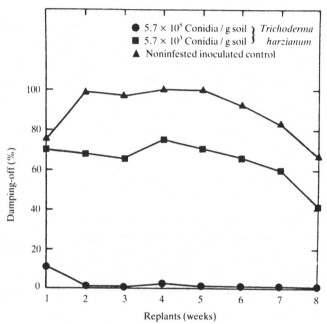

Table 7.1 *Virulence of healthy (189HTS) and hypovirulent (diseased, 189a) isolates of* R. solani *against sugar beet*

Isolate	Sugar beet seedlings			Postemergence damping off (%)
	Emerged (no.)	Killed (no.)	Healthy (no.)	
None	1866 y	1 y	1865 y	0 y
189a (diseased)	1850 y	2 y	1848 y	0 y
189a + 189 HTS	1872 y	52 y	1820 y	3 y
189 HTS (healthy)	1374 z	1085 z	289 z	79 z

Values in each column followed by different letters are significantly different, $P = 0.01$.
From Castanho, B. & Butler, E. E. (1978). *Phytopathology* **68**, 1511–14.

7.3   Hypovirulence

Hypovirulence has been reported by Castanho & Butler (1978) for *Rhizoctonia*. It seems to be similar to that for other pathogens (section 4.2.3) and involves the cytoplasmic inheritance of a factor that decreases the growth rate and the number of sclerotia produced, as well as reducing virulence. The hypovirulent strains possess extra dsRNA, though no virus-like particles have been seen. Unfortunately it seems to be rather difficult to transfer the hypovirulence, even within the same anastomosis group, so healthy strains are not easily infected when grown with the hypovirulent ones and there are at least eight anastomosis groups between which transfer is not possible, because *Rhizoctonia* does not produce asexual spores. When co-inoculated with some virulent strains the hypovirulent ones do reduce disease (Table 7.1) but this has not been extensively used in biocontrol.

7.4   The use of *Trichoderma*

T. hamatum, *T. harzianum* and to a lesser extent *T. koningii*, and *T. viride* have all been used against damping-off caused by *Rhizoctonia* and *Pythium* in the laboratory, the glasshouse and the field (Papavizas, 1985). Commercial preparations are available, though not very widely used. Isolates are obtained from suppressive soils (see above) and from a great variety of screening and selection procedures. Depending upon the particular isolate, control may be by the production of antibiotics or by mycoparasitism.

Simply adding *T. hamatum* to soil can reduce the damping-off of peas caused by *Pythium* (Fig. 7.2). *T. harzianum* and *T. koningii* will also do this, but they are not effective against *Rhizoctonia*. Other workers, with other isolates, have shown effective control of *Rhizoctonia* and *Sclerotium rolfsii* with *T. hamatum*. Such control can be just as effective as the conventional fungicides and may last much longer, by establishing in the soil and surviving until the next crop (Fig. 7.3).

The form of the inoculum is critical and it may be necessary to provide some nutrients to enable the conidia of *Trichoderma* to germinate, grow and sporulate to increase the effective inoculum levels. Adding the antagonist as young cultures grown on ground up wheat bran produced the most effective control in a series of experiments by Lewis and co-workers (Table 7.2). Control of *Rhizoctonia* by two different strains of *T. hamatum* was only effective from growing mycelial bran cultures which gave antagonist populations four to seven orders of magnitude greater than mycelium with no food source or conidial preparations with or without a food source. There are, however, complications because other antagonists may not be best from bran cultures. Thus, in this study, *T. hamatum* still gives much better control from mycelium in bran (Table 7.3) but *T. viride* and *T. harzianum* do not. The bran alone also makes the damping off slightly, but not significantly, worse. There are other reports where it seems the bran can provide a food source for which the pathogens (*Rhizoctonia*, *Phytophthora* and *Pythium*) successfully compete, so making the disease worse. There were problems in earlier work with contamination of the inoculum by a variety of saprotrophs exploiting the food source and making the antagonist ineffective. Additional studies on these *Trichoderma* isolates (and *Gliocladium*) have shown that pelleting the biomass and bran with alginate can increase possible storage time to months. There seems to be little relationship between the quantity of antagonist in the preparation and the effectiveness of the control, provided a minimum value is exceeded. So care is needed in inoculum formulation and different forms may be necessary for different antagonists or with different pathogens.

The antagonists may have different temperature optima from each other and from the pathogens. There is thus a complicated interplay depending on the environmental temperature. Damping off of cotton can be controlled in the glasshouse (27°C) by *T. harzianum* but at 20°C in the field *T. hamatum* worked best and gave as good control as the conventional pentachloronitrobenzene (PCNB) fungicide. *T. har-*

Fig. 7.2. Effect of *Trichoderma hamatum* added to a conducive soil at 10^6 propagules/g on reduction of damping-off of peas over time, in soil naturally infested with *Pythium* spp. pathogenic to peas. (From Chet, I. & Baker, R. (1981). *Phytopathology* **71**, 286–90.)

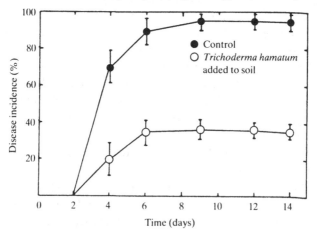

zianum also works against *Sclerotium rolfsii* and *R. solani* attacking beans in the field when the temperatures are above 20°C, and the control is best in more acidic soils.

There are also interactions between these antagonists and other soil or seed surface flora and fauna. There is selective grazing by some collembola (small soil arthropods) that fortunately prefer to eat the *Rhizoctonia* rather than the antagonistic *Trichoderma* on cotton seeds, so that they help in the biocontrol.

In another study *Pseudomonas* itself was a problem. A supposed biocontrol strain of *Pseudomonas* gave a very slight amount of control

Fig. 7.3. Percentage of healthy pea or radish seedlings at various intervals after planting. Seedlings grew from non-treated seeds in soil in which treated seeds of the same species had been sown before. Peas were planted in soil containing indigenous *Pythium* spp., while radish were planted in similar soil to which *Rhizoctonia* had been added. Bars labelled SD represent the overall standard deviation for the experiment. Captan, Dexon and PCNB (pentachloronitrobenzene) are commercial fungicides. (From Harman, G. E., Chet, I. & Baker, R. (1980). *Phytopathology* **70**, 1167–72.)

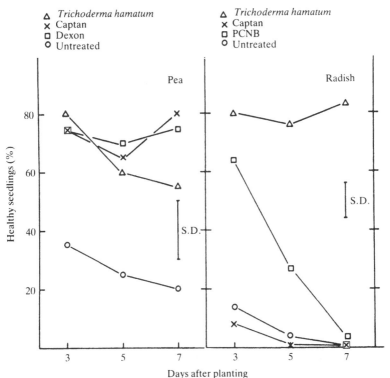

of *Pythium* on pea seeds, but was deleterious in antagonizing the *T. hamatum* and significantly reducing the effect of this latter agent. This interaction is most noticeable in soils low in available iron where the siderophores from the pseudomonad are most effective. This antagonism between antagonists is known from a number of applications, though it is not widely published. Researchers often take the 'best' strains of a biocontrol agent(s) and try mixed inocula, but usually there is not an additional effect of the combined antagonists. It may ultimately be possible to use mixtures to increase the range of environmental

Table 7.2 *Effect of preparations of isolates of* Trichoderma hamatum *on survival and saprophytic growth of* Rhizoctonia solani *and on antagonist proliferation in soils. Conidia were added directly to soil at* 5×10^3/g *of soil. Conidial preparations were conidia on bran mixed immediately with soil (1:200 w/w) to give an antagonist density of* 5×10^3 *conidia/g of soil. Mycelium was grown in potato dextrose broth for 2 days and added to the soil at* 5×10^3 *propagules/g. Mycelium preparations were 3 day old cultures of the antagonist on bran, added to soil (1:200 w/w) to give an antagonist density of* 5×10^3 *propagules/g*

| Isolate and preparation | *R. solani* (R–23) | | |
	Survival in infested beet seed (%)	Saprophytic growth (% beet seed colonization in soil)	Antagonist population (colony-forming units g soil)
Control	97 a	89 a	3×10^3
Bran	93 a	92 a	6×10^3
T. hamatum			
Tm–23			
Conidia	97 a	92 a	5×10^2
Conidial preparation	96 a	95 a	1×10^3
Mycelium	94 a	83 a	4×10^3
Mycelial preparation	33 b	0 b	2×10^8
TRI–4			
Conidia	95 a	94 a	2×10^3
Conidial preparation	98 a	90 a	6×10^2
Mycelium	88 a	78 a	2×10^5
Mycelial preparation	2 b	0 b	4×10^9

Numbers in each column for each isolate followed by different letters are significantly different from each other, $P = 0.05$.
From Lewis, J. A. & Papavizas, G. C. (1985). *Phytopathology* **75**, 812–17.

factors over which control will operate, but too little is known about colonization and competition with indigenous flora for a single antagonist to predict the effect of mixtures.

These environmental and formulation interactions most probably explain the variable results sometimes obtained in the use of *Trichoderma* against damping off; commercially the control by fungicides is still preferred as being more reliable. Integrated control, using both biocontrol and chemicals, could be one of the ways around the difficulty. It has been known for some time that doses of PCNB as low as 1–2 µg/kg soil, which are not sufficient to be lethal, can be effective in combination with *Trichoderma*. This implies that the antagonist is more resistant to the fungicide than the pathogen and/or that the slightly debilitated pathogen is more easily attacked than when unaffected by fungicide. A more positive development of this strategy has been the production of special strains of *Trichoderma* which are fungicide tolerant. Benomyl can be used in the treatment of *Rhizoctonia*, though

Table 7.3 *Effect of mycelial and conidial preparations of isolates of* Trichoderma viride *(T–1–R4),* T. harzianum *(WT–6–24),* T. hamatum *(TRI–4) and* Gliocladium virens *(GI–21) on cotton, sugar beet and radish seedling stands in soil infested with* Rhizoctonia solani

Isolate and preparation	Plant stand (%) at 3 wk		
	Cotton	Sugar beet	Radish
Control (noninfested)	96 a	53 a	76 a
R. solani	2 b	8 c	3 c
R. solani and bran	6 b	13 bc	8 bc
T. viride (T–1–R4)			
Mycelial preparation	25 b	21 b	31 b
Conidial preparation	18 b	16 bc	9 bc
T. harzianum (WT–6–24)			
Mycelial preparation	18 b	11 bc	14 bc
Conidial preparation	16 b	21 b	29 bc
T. hamatum (TRI–4)			
Mycelial preparation	86 a	63 a	70 a
Conidial preparation	10 b	8 c	15 bc
G. virens (GI–21)			
Mycelial preparation	94 a	70 a	83 a
Conidial preparation	20 b	21 b	38 b

Details as in Table 7.2.

there are problems of resistance to the fungicide (see also section 3.3). *Trichoderma* isolates have been exposed to UV light and mutants selected for enhanced inhibition of *Rhizoctonia* and tolerance of benomyl, captan and pentachloronitrobenzene. The selected isolates give better control of damping off than the wild type. A similar study with rotting of carnation cuttings by *Rhizoctonia* showed that an isolate of *T. harzianum* reduced rotting by 50% (Fig. 7.4) and benomyl gave rather better control, depending on the inoculum density of the pathogen. Both were applied in the rooting hormone powder used in the propagation of the cuttings. When a benomyl tolerant isolate was used, together with the fungicide, there was no rotting at all (Fig. 7.4). These integrated control measures mean that fewer applications of fungicide can be used, so reducing the environmental impact, or alternatively control can be obtained in situations where neither the biocontrol agent nor the chemical were completely effective. Integrated control may combine some of the dependability of chemicals with the environmental desirability of biological control.

Fig. 7.4. Integrated chemical and biological control of *Rhizoctonia* stem rot during propagation of carnation cuttings. The T–95 is an isolate of *Trichoderma harzianum* mutated so that it was tolerant to benomyl. Benomyl (50% a.i.) was applied as a 5% suspension. (From Baker, R. & Scher, F. M. (1987). In *Innovative approaches to plant disease control*, ed. I. Chet. New York: Wiley.)

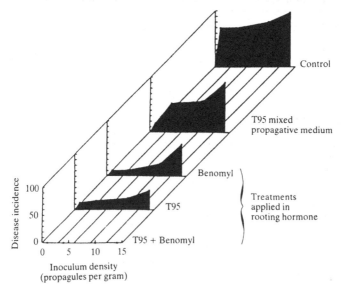

7.5 Other antagonists against damping off

The use of many organisms as seed inoculants has been reviewed by Kommedahl & Windels (1981), but *Gliocladium* is the only other fungus which has been widely tested against damping off and its development has gone side by side with that of *Trichoderma* to which it is closely related. It is a mycoparasite and also produces antibiotics. In this case the best inoculum formulation was a conidial preparation (Table 7.3), rather than mycelium. When used as a peat based inoculum on cotton, *Gliocladium* more than halved the pre-emergent damping off by both *Rhizoctonia* and *Pythium* but was less effective against post-emergent damping off.

Another fungus whose use has been suggested for the control of pre-emergent damping off is *Pythium oligandrum*. The colonization of seeds by *P. ultimum* was reduced from 77 to 10% by competitive exclusion by *P. oligandrum*, which can also act as a mycoparasite. It is now available commercially on a small scale.

Of the possible bacterial inoculants *Bacillus subtilis* is the most promising and has been investigated for the control of seedling blight of maize where it gave as good control as the conventional captan or thiram fungicides, provided that the soil moisture levels were quite high. *Enterobacter cloacae* is a normal inhabitant of seed coats (in particular of beet, peas and cucumber) which increases during normal germination, protecting the seedlings from *Pythium*. If artificially inoculated onto seeds sown in soil infested with *Pythium* it doubled the survival

Table 7.4 *Deleterious effect of protozoa (2 amoebae Am1 and Am2, and a ciliate Cil) on seed bacterization with a plant growth promoting Pseudomonas (Ps)*

Treatment	Dry weight		Kj–N	
	g	% of control	% on DW	Total amount (mg) per treatment
Control	1.256 a	100.0	2.15 a	136.1
Ps	1.509 c	124.2	2.16 ab	162.3
Ps + Am1	1.407 bc	115.8	2.44 c	171.7
Ps + Am2	1.323 ab	108.8	2.39 bc	158.1
Ps + Cil	1.267 a	104.2	2.69 d	170.4

Results followed by a different letter are significantly different at $P = 0.05$.
From Vandenabeele, J. & Verstreate, W. (1986). *Med. Fac. Landouww. Rijksuniv., Gent.* **51**/3b, 1363–69.

rate. This bacterium has the advantage that it is a normal inhabitant of seeds, and of the human gut, so may be expected to raise no serious safety problems (see section 6.7). Neither of these bacteria is available commercially for this application.

Pseudomonas has, of course, also been tried against damping off and seedling diseases, but has not been so successful here. Not only can it antagonize other control agents (see above), but it has itself been shown to be consumed by protozoa. The effect of protozoa grazing on biocontrol agents is complicated. Bacterial inoculants are frequently added, or exist in quite high numbers and could be suitable food for bacterivorous ciliates and amoebae. A seed inoculated *Pseudomonas* increased plant dry weight but made no difference to nitrogen content (N%, Table 7.4). When amoebae (Am1 and 2) and especially ciliates were added, the growth increase caused by the *Pseudomonas* was reduced but the nitrogen content of the plants increased as protozoan predation mobilized the nitrogen in the bacteria, an effect known from general soil studies as well as biocontrol. It may be that this predation on potential control agents occurs in bulk soil too, but it has only been shown for seedling diseases.

7.6   Conclusion

The use of biological control against damping off and related diseases has promise. There are commercially available inocula in some countries, though this is not widely used for control. There are many more organisms that clearly give protection and could be developed. These 'new' systems have not proved to be commercially competitive with the well-known, cheap, effective and reliable fungicides which the grower is accustomed to use. However, the intensive horticultural systems, often with specialized, more or less sterile composts and controlled environments, make this a potentially useful site for biological control.

8

Conclusions and perspectives

Let us try to draw together some of the conclusions of the previous chapters and, what is much more difficult, try to predict where the studies will go in the next few years.

It is quite clear that there are more than enough examples of laboratory demonstrations with different diseases in different parts of the plant with different possible control agents. That is not to say that all possible control agents have been found, indeed we have hardly started looking. Even though it is possible to make a short list of the most likely organisms (*Trichoderma*, *Pseudomonas*, *Bacillus*, for example) that should not discourage the search for particular species or strains for the environment, plant or disease under study.

What is lacking is basic information on the environmental and genetically determined factors that control survival, colonization and effectiveness in the field. We need to know what morphological and physiological characteristics of an organism will make it successful in a particular environment. Should an organism be motile? Should it be able to exist in low nutrient conditions? Is there some special character(s) that means an organism is fitted for growth on the root, or better the root of a particular plant host? We just do not know. There is talk of using genetically engineered organisms, and we have seen examples of some of these. Maybe the molecular biologists can put in this character or take out that: such techniques are no good to us until we know what to ask for. If we could say 'put in such and such an enzyme' or 'make the wall thicker or thinner or take out the polysaccharide' then we could make a start. We may now be able to make some minor adjustments by genetic engineering or selection to allow us to track an organism in the environment, or to improve its storage or fermentation properties, but we cannot even start to 'design' an organism for a particular job. We may in future be able to breed hosts to favour particular antagonists or alternatively antagonists for particular hosts.

For the foreseeable future the basic process of selection will be random, perhaps a little biased by knowledge. This may not be altogether a bad thing: it is essentially the process used to select chemical control agents. This has produced many excellent compounds over the last 40 years, though admittedly also some compounds which, with hindsight, ranged from unsuitable to definitely damaging to man or the environment.

We can improve the selection techniques by using live plant tests rather than *in vitro* assays where possible. There are however basic problems with the methodology of tracking organisms released into environments. A lot of work is going on with this; it is a priority in the centrally funded research of many countries and in the large commercial firms. There are a few tricks to help with some organisms, like using bait plants to find *Rhizobium* which are host plant specific, but the general saprotrophs are dependent on selective media in dilution plating systems. Various immunological methods and DNA matching systems are under test to identify particular strains of organisms, especially bacteria, but almost all of these require a culture first, so you are back to dilution plates or most probable number methods. The problem of environmental protection is serious, or has been perceived as such by many organizations and the general public. It is not insuperable and the release of genetically engineered or selected organisms will eventually be allowed, though with proper care and legislative controls.

There is a good deal of information and expertise on fermenter technology, production of inoculum and formulation. Not all is known by any means, and each organism presents its own challenge, but this stage is not the limiting step in most biological control programmes. There will no doubt be improvements when there are real problems to be solved, but there seems no sign of insuperable difficulties. Seed coating seems to be the delivery system of choice for soil-borne diseases and again there are lots of ideas about. There will be further development of this as needed, but the main consideration is to get the inoculum to the site of action, with as little spread round the general environment as possible. This site may be defined places like tree wounds or may be resting structures of a fungus in the bulk soil, so the problems are specific to each disease.

There are still more than enough diseases with no good chemical control and no host resistance, so there are still plenty of openings for biological and integrated control without competing directly with chemicals which are already used.

Successful examples of biological control are from general agricultural methods and amendments which work microbiologically in traditional agricultural systems, and from the introduction of antagonists to more or less empty environments such as tree stems and stumps, nursery composts or biological voids created artificially by the use of soil fumigation. The next main target diseases are glasshouse and horticultural crops and soil-borne diseases of field crops. There are beginning to be signs of success though there is still only minimal investment of time and money compared with chemical control. The most likely developments will be in varying degrees of integrated control, combining cultural practices (rotation, tillage, host cultivar selection etc.) with the application of antagonists and the management of the crop to favour the introduced organisms (if only we knew what they needed!).

Finally, to end on an optimistic note, there is a future for biological control but the routine treatment of many field diseases is still a long way off. The recent increase in research effort and funding will not bear fruit for at least 10 years in terms of widely available commercial inoculum. Biological control should take less time, and cost less money, than the present sophisticated chemical control systems have involved. New research must be centred on realistic agricultural and horticultural situations, rather than on laboratory demonstrations. Commercial financial considerations and the restrictions or encouragements from environmental protection agencies and central government or supra-governmental agencies (such as the European Economic Community and its agricultural policy) will be crucial in determining the economic climate and the public opinion within which biological and integrated control will have to operate. We have produced, and will in future produce, biologically viable systems but their widespread use will depend on these policies outside the direct control of the scientists developing the systems.

Glossary

This glossary has been compiled from definitions and discussions in the text and also owes much to standard reference books including *A guide to the use of terms in plant pathology* (Commonwealth Mycological Institute, Kew, England. Phytopathological Papers 17. 1973), *Dictionary of the fungi* (Edited by Ainsworth, G. C. 1971. Commonwealth Mycological Institute, Kew. 2nd edition), and *Dictionary of microbiology* (Singleton, P. & Sainsbury, D. (1978). Chichester: John Wiley).

cf. (Latin, *confer*), means compare the definition of the word indicated with the definition being considered.

q.v. (Latin, *quod vide*), means that the word is defined elsewhere in the glossary.

aerated steam Mixture of air and steam to produce defined temperatures when injected into soil so that particular components of the flora can be killed: e.g. kill the pathogens and leave the spore-forming saprotrophs to recolonize the soil. *cf.* soil sterilization.

aflatoxins A group of toxins *q.v.* produced by fungi (especially *Aspergillus flavus*) growing on seeds: they affect animals eating the seeds and may cause death.

amensalism A form of symbiosis *q.v.* in which one organism is harmed by another: e.g. one produces an antibiotic. The basis of much microbial antagonism *q.v.*

antagonist (hence **antagonism**) An organism exerting a damaging effect on another: e.g. by the production of lytic enzymes (cf. lysis) or antibiotics *q.v.*, or by competition *q.v. cf.* specific and general antagonism.

antibiotic A substance produced by a micro-organism which is damaging to another at low concentrations (μg/ml). *cf.* toxin, bacteriocin.

arbuscules Finely branched fungal hyphae within the cell (outside the plasmalemma) of a plant root which are part of some mycorrhizal *q.v.* associations. From Latin, *arbor* a tree in reference to the general shape.

autotroph An organism which does not require reduced (usually organic) carbon for its energy source. Autotrophs usually use light energy (photosynthesis) but some bacteria can use reduced inorganic compounds as an energy source.

avirulence The lack of ability to cause disease. *cf.* virulence. A strain *q.v.* of a pathogen *q.v.* may be avirulent, even though the species or genus as a whole are pathogens.

bacteriocins Chemicals, usually proteins, which are produced by micro-organisms and which act as antibiotics *q.v.* against closely related species or strains of micro-organism.

binding sites Regions of the cell wall of an organism which contain particular chemicals (usually proteins, polysaccharides, glycoproteins or lipopolysaccharides) to which other organisms can specifically bind by the formation of chemical bonds. These are often the basis of recognition between a host and a pathogen and may be partly responsible for the specificity of the interaction.

biocide A chemical which kills a wide spectrum of organisms: a general poison. *cf.* pesticide.

biotroph A pathogen that requires living host tissue for its nutrition. *cf.* necrotroph.

canker An area of necrotic *q.v.* tissue, especially in a stem, caused by a pathogen. There may be disturbance of the growth of the stem producing swelling around the dead tissue.

combative Competitive organisms which defend a captured resource *q.v.* See Fig. 1.3.

commensalism A form of symbiosis *q.v.* in which one organism benefits another, without itself being harmed or helped: e.g. the production of a vitamin which allows growth of other organisms in a micro-environment.

community The micro-organisms, of several different taxa, which live together in a habitat *q.v.* A collection of populations *q.v.*

competition An interaction between organisms which both need a limited resource e.g. a food supply, so that the preferential use of it by one organism harms the other or reduces its growth rate.

competitive exclusion One organism is so much better at acquiring a limited essential resource that it effectively excludes all others from the microhabitat.

competitive saprophytic (saprotrophic) ability The capacity of a pathogen to compete with saprotrophs *q.v.* during a stage in its life cycle when it is not being pathogenic.

compost (a) The result of controlled decomposition of organic matter to produce a material, with reduced C/N ratio, suitable for adding to the soil.

(b) A general term for the mixtures of sand, soil and organic matter (e.g. peat) to form an artificial growing medium, especially for seedling and horticultural crops.

conducive soil A soil which allows development of disease: the opposite of suppressive *q.v.*

cross-protection The protection of a host from disease produced by inoculating a strain *q.v.* or isolate *q.v.* closely related to the pathogen: e.g. a different *forma specialis q.v.* or an avirulent *q.v.* strain. Protection may be by competition *q.v.* etc., or by induced resistance *q.v.*

cultivar A subdivision of a species, usually of a plant, which has been produced by an artificial breeding programme on a plant in cultivation. *cf.* variety.

deleterious bacteria Bacteria, usually growing in the rhizosphere *q.v.* which reduce plant growth: e.g. by the production of cyanide, without obvious symptoms of disease. *cf.* minor pathogens.

delivery system The means by which an antagonist (or chemical) is applied to the site of action: e.g. as a spray, a dry granule, a seed coat, etc. *cf.* formulation.

direct drilling Sowing seeds of a crop directly into the remains of the last one without any cultivation. The same as no-till. *cf.* minimum tillage.

disease decline The reduction in disease during monoculture *q.v.* Initially there may be an increase in disease severity as inoculum of the pathogen builds up, but after a few years the disease declines to a low level: e.g. take-all decline. The soil becomes suppressive *q.v.*

drench A pesticide *q.v.* which is applied in a large volume of water to saturate a plant or especially the soil.

dsRNA Double stranded ribose nucleic acid. It may be transferred in the cytoplasm when vegetatively compatible *q.v.* mycelia fuse and may cause hypovirulence *q.v.*

edaphic Concerned with soil.

elicitor A chemical which initiates a response to infection by the host, especially a fungal product which causes the production of a phytoalexin *q.v.*

environment All the biotic and abiotic factors which make up the surroundings of an organism, hence **micro-environment**.

epidemiology The study of the spread of disease, particularly of animals. The spread of a plant disease should be called an epiphytotic (equivalent to epidemic), but the terminology is little used.

forma specialis A subdivision of a species, usually used of a pathogen, where different races *q.v.* are specific to different host species, while being the same morphologically. The abbreviation is *f. sp. cf.* variety, strain.

formulation The mixture, and means of preparing it, by which an antagonist (or chemical) is packaged, stored and delivered to the point of use. It may include a food base, wetting agents, inert diluting agents, water binding or desiccating agents, etc. as well as the active ingredient or actual micro-organisms.

fumigant A pesticide *q.v.* which operates in the vapour phase, especially in soil 'sterilization' *q.v.*

fungicide A chemical which kills or harms a fungus. *cf.* pesticide.

fungistasis The prevention of fungal growth, mainly by carbon limitation. The production of germination inhibitors by other organisms, or by the organism itself, may also be included.

general antagonism Reduction in disease or pathogen inoculum due to general microbial activity: e.g. on adding available organic matter to soil. *cf.* specific antagonism.

general suppression The characteristic of a suppressive soil *q.v.* which is due to the overall structure and microbiology: e.g. clay content and general microbial activity on adding organic matter. *cf.* specific suppression.

genetic engineering The artificial manipulation of the genome of an organism to produce strains *q.v.* or cultivars *q.v.* with characters not previously occurring or not occurring in that combination before. It especially refers to the recent ability to change genomes by the transfer of DNA.

germination lysis The lysis *q.v.* of spores or germlings by micro-organisms after stimulation of germination by available nutrients or specific chemicals.

gnotobiotic Pertaining to the culture of known combinations of organisms: e.g. a plant with no associated microbes or with only a known strain, or strains in known combinations.

green manure Incorporation into soil of fresh organic material, other than just plant residues, that has been grown either *in situ* or elsewhere.

habitat The place where an organism lives. *cf.* niche.

hyperparasitism See mycoparasitism.

hypovirulent Of reduced virulence *q.v.*, especially of strains or races of a pathogen infected with a virus or dsRNA *q.v.*

iatrogenic Diseases caused by man's activities e.g. a fungicide application making a disease worse when the pathogen is resistant to the fungicide and the competing saprotrophs are not.

ice nucleation The formation of ice crystals on ice nuclei that may be contained in general dust etc. or may be specific proteins produced by bacteria. In the presence of ice nuclei water crystals form near to 0 °C, and so damage the plants. Without ice nuclei water may supercool to −5 °C or even −10 °C without forming ice and therefore without damaging the plant.

immune Complete freedom from infection by a pathogen because of some characteristic of the host, hence **immunity**. *cf.* resistance.

in vitro A test or procedure carried out (literally 'in glass') in laboratory conditions: e.g. testing an antagonist in a petri dish against a culture of the pathogen.

in vivo A test or procedure carried out with live host material: e.g. testing an antagonist against a pathogen that is growing on or in its living host.

induced resistance A form of cross-protection *q.v.* in which the increased resistance of the host is caused by stimulation of the host defence systems after inoculation with an isolate of the pathogen which is avirulent *q.v.* or which does not cause disease on that host: e.g. a different *forma specialis q.v.*

inoculate Introduce a living organism (the **inoculum**) into a culture medium or the environment.

inoculum (inoculant) An organism introduced into an environment, usually after being produced artificially in the laboratory, to carry out some special function: e.g. disease control, plant growth promotion, straw decomposition, etc.

inoculum potential The sum of all the factors that contribute to the energy available to a pathogen for infection of the host: e.g. inoculum particle size, food base, inoculum density, total amount of inoculum, etc.

insecticide A chemical which kills or harms an insect. *cf.* pesticide.

integrated control A combination of chemical, biological and agricultural practices which together control a disease. An overall approach to disease control.

K-strategists Organisms with low reproductive rates but which are competitive: e.g. producing enzymes to degrade complex polymers. Population sizes relatively stable. See Fig. 1.3. *cf. r*-strategists. '*K*' stands for the carrying capacity of the environment in equations on population dynamics.

lysis The rupture or dissolution of a cell, hence lytic. Endolysis (= autolysis) is the breakdown of the cell by its own enzymes following death which may be caused by starvation, antibiotics etc. Exolysis (= heterolysis) is the destruction of the cell by enzymes of another organism: e.g. in mycoparasitism *q.v.*

microbial interactions See competition, neutralism, symbiosis, predation, parasitism, commensalism, amensalism, antagonism.

microhabitat A small habitat *q.v.*, often only micrometres in dimensions, where a micro-organism lives.

minimum tillage A form of cultivation where something less than ploughing is done: e.g. just rotovating the surface. The ultimate in minimum tillage is no-till or direct drilling *q.v. cf.* tillage.

minor pathogen Micro-organisms, often several at once, which produce disease symptoms, but are usually so inconspicuous that they are not noticed, and are only shown to occur when their effect is removed by soil sterilization *q.v. cf.* deleterious bacteria.

mixed cropping The growth of more than one species of crop on the same ground at the same time: e.g. wheat undersown with legume.

monoculture The growing of the same crop on the same ground year after year. *cf.* rotation.

mutualism A form of symbiosis *q.v.* in which both organisms derive some benefit, a mutually benficial relationship: e.g. a legume root nodule, a mycorrhiza, a lichen.

mycoparasitism The parasitism of one fungus by another (= hyperparasitism).

mycorrhiza A symbiotic *q.v.*, mutualistic *q.v.* relationship between a root and a fungus. Many different types of polyphyletic origin (i.e. have arisen many times during evolution).

mycotoxins Toxins *q.v.* produced by fungi, usually growing on seeds and fruits. Animals which subsequently eat the seeds are affected by the toxins.

necrosis Death, usually of a limited part of a plant, associated with a disease, hence necrotic.

necrotroph A pathogen which kills host cells and then lives on the dead tissue. *cf.* biotroph.

neutralism A form of symbiosis *q.v.* in which, though the organisms live together (e.g. in the same microhabitat) they do not affect each other, they occupy different and non-overlapping niches *q.v.* in the same habitat *q.v.*

niche The sum of all the physiological properties of an organism, of the environment and its exploitable resources which together define what a microbe does in that habitat *q.v.* Fundamental niche: the niche which an organism is inherently capable of utilizing. Realized niche: the actual niche available because of biotic or environmental limitations.

no-till See direct drilling.

nutrient limitation The normal condition in most natural habitats (soil, leaves, plant stems, etc.) in which microbial growth is restricted or stopped by a lack of available nutrients.

opportunistic pathogens Ones that will take advantage of a weakened host to cause disease. The organisms may normally be saprotrophs *q.v.* not able to overcome the host's defences, unless the host is injured or incapacitated by another disease or adverse environmental conditions.

organic amendment The addition of organic matter to soil.

organic farming Growing crops and animals without the use of chemically synthesized fertilizers, pesticides, feed additives, etc. Includes the use of manures, crop residues and rotation of crops to maintain soil fertility and to control weeds, pests and diseases. Uses biological control *sensu lato*.

parasite An organism living on or in another living organism (the host) from which it obtains its food. If it harms the host in the process (causes a disease) it is a pathogen *q.v.*, but not all parasites are pathogens. See symbiosis.

pathogen An organism capable of causing disease in a plant, animal or microorganism. *cf.* parasite. It may apply in general to genera, species or strains, though particular isolates of the species or strain may vary in virulence *q.v.*

pathogenicity The tendency of a member of a group of organisms known to be pathogens *q.v.* to actually cause disease. *cf.* virulence, resistance.

pesticide A chemical which kills or harms a pest. May be divided according to which group of animals it is toxic to: e.g. insecticide, molluscicide, nematocide, etc. Also used as a general term to include all biocides *q.v.* used against pathogens and pests, i.e. including fungicides.

phyllosphere The environment around a leaf which is influenced by that leaf. Hence **phylloplane**, the actual leaf surface. *cf.* rhizosphere.

phytoalexin A chemical produced by the host plant in response to infection, which is toxic to fungi and bacteria.

plant growth promoting rhizobacteria (PGPR) Bacteria growing in the rhizosphere *q.v.* that increase the growth or yield of a plant by controlling minor pathogens *q.v.* or deleterious bacteria *q.v.*, or possibly by affecting the amount and distribution of plant hormones.

plasmid An extrachromosomal piece of DNA, especially in bacteria, that can replicate independently of the main chromosome and may be transferred between cells.

population A group of similar organisms, usually a single species if this can be defined, that live in the same habitat (= a colony). Many different populations make up a community *q.v.*

predation A form of symbiosis *q.v.* in which one organism eats another, hence **predator.**

race A subdivision of a species which is genetically distinct: e.g. races of a pathogen may have different virulence *q.v.* genes. Also a geographically distinct mating group. *cf.* variety, strain, *forma specialis*.

resistance To disease: the ability of a host to overcome, to some extent anyway, the invasion of or symptoms produced by a pathogen *q.v. cf.* virulence, immunity.
To agrochemicals: the loss of effect of a fungicide, for example, as the pathogen mutates to overcome the toxicity: the fungus is resistant to that fungicide.

resource A component of the environment that can be utilized by an organism, usually as food. **Primary resource**: originating from a primary producer: e.g. a part of a plant. **Secondary resource**: originating from a consumer: e.g. a fungal cell wall which had grown on the plant.

resource capture The occupation of a resource, followed by the exclusion of other organisms. *cf.* competitive exclusion, combative organisms.

resource competition See competition and resource capture.

***r*-strategists** Organisms with a high reproductive rate, common in situations with abundant food (= *r*-selected organisms). Populations of active *r*-strategists fluctuate greatly. See Fig. 1.3. *cf. K*-strategists. '*r*' stands for the rate of population increase in equations for population dynamics.

rhizosphere The volume of soil around a root, under the influence of that root, in which microbial activity is increased. **Rhizoplane**: the root surface, but this is generally very difficult to find except in young roots, as epidermal and cortical cells die and are colonized by micro-organisms. Hence a more restricted definition, **endorhizosphere**: the outer layers of the root cortex which are colonized by micro-organism, **ectorhizosphere**: the rhizosphere outside the root.

rotation Growing a repeated sequence of different crops on the same ground in successive years, usually in a cycle of three to six years: e.g. wheat, wheat, grass then wheat, wheat, grass again. *cf.* monoculture.

ruderal Organisms characteristic of disturbed environments, and hence short lived, good at resource capture *q.v.*, often *r*-strategists *q.v.* See Fig. 1.3.

saprophyte An organism (plant or microbe) which lives on dead organic material.

saprotroph An organism which depends on non-living material for its nutrients.

sclerotium A resting structure of fungi composed of a mass of hyphae, usually surrounded by a waterproof layer with thickened hyphal walls.

screen A test, involving large numbers of repeats of a standard procedure, the object of which is to find a few useful organisms, or chemicals, from the general population. A primary screen may be very large, a secondary screen may further test the potentially useful survivors from the primary screen.

siderophore A chemical, produced by an organism, which binds cations, especially Fe^{3+} and helps to transport it into the organism in iron limited environments.

soil sterilization The treatment of soil with a biocide *q.v.* or heat to kill all organisms. Usually less rigorously used, for commercial treatment is not often severe enough to kill everything, but will be designed to at least kill the pathogens.

solarization The use of the sun's heat for soil sterilization *q.v.* The soil is covered with black polythene to increase absorption of radiation and to conserve heat.

specific antagonism Reduction in disease or in pathogen inoculum due to the activity of a particular species or strain of micro-organism. *cf.* general antagonism.

specific suppression The characteristic of a suppressive soil *q.v.* which is due to a single antagonistic organism or a closely related group of antagonists (which may be natural or artificially inoculated). *cf.* general suppression.

strain A group of similar, characterized, isolates *q.v.* of a micro-organism. Essentially this applies to laboratory isolates, cultures or selections. *cf.* race, variety, *forma specialis.*

strategy The general way in which a micro-organism lives; the strategy may be to produce many spores so that there is always one ready should a food source arrive, or to use a food that no other organism has the enzymes to degrade etc. *cf.* *r*-strategist, *K*-strategist, ruderal, stress tolerant, combative.

stress tolerant Organisms which persist in stressed environments: e.g. low food, low water, high temperature. See Fig. 1.3.

suppressive soil A soil in which disease is reduced or absent, even though the pathogen is present or introduced and a susceptible host is grown. The opposite of conducive *q.v.* *cf.* general and specific suppression, decline.

symbiosis The living together of two or more organisms, regardless of the nature of the interdependence, hence **symbiotic**. *cf.* mutualism, amensalism, commensalism, neutralism, parasite, pathogen, predator, synergism.

synergism A form of symbiosis in which the organisms exchange metabolites or have complementary enzyme systems: e.g. in the breakdown of the various polymers in a cell wall, but the relationship is not obligatory.

tillage Cultivation of ground before sowing seed or during fallow. Hence no-till systems sow seeds into the remains of the last crop without ploughing etc. *cf.* minimum tillage.

toxin Any substance, usually organic, which is harmful to an organism, a poison. *cf.* antibiotic.

tylosis A blockage of the conducting vessels of a plant by the expansion into them of neighbouring live cells through pits in the wall of the conducting vessel.

variety A subdivision of a species of plant or animal, below the rank of sub-species. Not necessarily produced by a breeding programme. *cf.* cultivar. Also used of pathogens as a subdivision of a species that may show some host specificity but is also distinct morphologically. Abbreviation is var. *cf.* strain, *forma specialis.*

vegetative compatibility The ability of different strains or races *q.v.*, especially of the mycelium of a fungus, to fuse and exchange nuclei. Incompatibility of closely related strains or races ensures outbreeding.

virulence The relative capacity of a pathogen to attack a host. A pathogen *q.v.* is virulent if it causes disease and overcomes host resistance *q.v.* *cf.* avirulent.

wild type The organism as growing in the natural environment, i.e. before laboratory culture, selection, mutation, etc.

References

Chapter 1

Andrews, J. H. (1984). Life history strategies of plant parasites. *Advances in Plant Pathology*, **2**, 105–30.

Andrews, J. H. & Harris, R. F. (1986). *r*- and *K*- selection and microbial ecology. *Advances in Microbial Ecology*, **9**, 99–147.

Bailey, J. A. (ed.) (1986). *Biology and molecular biology of plant-pathogen interactions*. Berlin: Springer-Verlag.

Bailey, J. A. & Mansfield, J. W. (1982). *Phytoalexins*. Glasgow: Blackie.

Baker, K. F. & Cook, R. J. (1974). *Biological control of plant pathogens*. San Francisco: Freeman.

Baker, R. (1968). Mechanisms of biological control of soil-borne pathogens. *Annual Review of Phytopathology*, **6**, 263–94.

Begon, M., Harper, J. L. & Townsend, C. R. (1986). *Ecology: individuals, populations and communities*. Oxford: Blackwell.

Blakeman, J. P. (ed.) (1981). *Microbial ecology of the phylloplane*. London: Academic Press.

Campbell, R. (1983). *Microbial ecology*. 2nd edition. Oxford: Blackwell.

Campbell, R. (1985). *Plant microbiology*. London: Edward Arnold.

Clapham, W. B. (1983). *Natural ecosystems*. 2nd edition. New York: Macmillan.

Cook, R. J. & Baker, K. F. (1983). *The nature and practice of biological control of plant pathogens*. St Paul, Minnesota: American Phytopathological Society.

Cooke, R. C. & Rayner, A. D. M. (1984). *Ecology of saprotrophic fungi*. London: Longman.

Elad, Y. (1986). Mechanisms of interactions between rhizosphere micro-organisms and soil-borne plant pathogens. In *Microbial communities in soil*, ed. V. Jensen, A. Kjøller & L. H. Sørensen. pp. 49–60. London and New York: Elsevier.

Francis, C. A. (ed.) (1986). *Multiple cropping systems*. New York: Macmillan Publishing Co.

Garrett, S. D. (1970). *Pathogenic root infecting fungi*. Cambridge: Cambridge University Press.

Gottlieb, D. (1976). The production and role of antibiotics in soil. *Journal of Antibiotics*, **29**, 987–1000.

Grime, J. P. (1977). Evidence for the existence of three primary strategies in plants and its relevance to ecological and evolutionary theory. *American Naturalist*, **111**, 1169–94.

Horsfall, J. G. & Cowling, E. B. (ed.) (1978). *Plant disease; an advanced treatise*. Vol. 2. *How disease develops in populations*. New York: Academic Press.

Horsfall, J. G. & Cowling, E. B. (ed.) (1980). *Plant disease; an advanced treatise*. Vol. 5. *How plants defend themselves*. New York: Academic Press.

Hutchinson, G. E. (1957). *A treatise on limnology*. Vol. 2. *Introduction to lake biology and the limnoplankton*. New York: John Wiley.

Ingold, C. T. (1978). Dispersal of microorganisms. In *Plant disease epidemiology*, ed. P. R. Scott & A. Bainbridge, pp. 11–23. Oxford: Blackwell.

Kosuge, T. & Nester, E. W. (ed.) (1984). *Plant-microbe interactions; molecular and genetic perspectives*. Vol. 1. New York: Macmillan Publishing Co.

Kuć, J. (1981). Multiple mechanisms, reaction rates and induced resistance in plants. In *Plant disease control*, ed. R. C. Staples & G. H. Toenniessen, pp. 259–72. New York: Wiley.

Lockwood, J. L. (1986). Soilborne plant pathogens: concepts and connections. *Phytopathology*, **76**, 20–7.

Lynch, J. M. (1983). *Soil biotechnology: microbiological factors in crop productivity*. Oxford: Blackwell.

Nelson, R. R. (ed.) (1973). *Breeding plants for disease resistance: concepts and applications*. University Park and London: Pennsylvania State University Press.

Palti, J. (1981). *Cultural practice and infectious crop diseases*. New York: Springer-Verlag.

Parker, C. A., Rovira, A. D., Moore, K. J., Wong, P. T. W. & Kollmorgen, J. F. (ed.) (1985). *Ecology and management of soilborne plant pathogens*. St Paul, Minnesota: American Phytopathological Society.

Schneider, R. W. (ed.) (1982). *Suppressive soils and plant disease*. St Paul, Minnesota: American Phytopathological Society.

Scott, P. R. & Bainbridge, A. (ed.) (1978). *Plant disease epidemiology*. Oxford: Blackwell.

Williams, S. T. & Vickers, J. C. (1986). The ecology of antibiotic production. *Microbial Ecology*, **12**, 43–52.

Windels, C. E. & Lindow, S. E. (ed.) (1985). *Biological control on the phylloplane*. St Paul, Minnesota: American Phytopathological Society.

Chapter 2

Alexander, M. (1986). Ecological concerns relative to genetically engineered microorganisms. In *Microbial communities in soil*, ed. V. Jensen, A. Kjøller & L. H. Sørensen, pp. 347–354. London: Elsevier.

Andrews, J. H. (1985). Strategies for selecting antagonistic microorganisms from the phylloplane. In *Biological control on the phylloplane*, ed. C. E. Windels & S. E. Lindow, pp. 31–44. St Paul, Minnesota: American Phytopathological Society.

Andrews, J. H. (1987). How to track a microbe. In *Microbiology of the phyllosphere*, ed. N. J. Fokkema & J. van den Heuvel, pp. 15–34. Cambridge: Cambridge University Press.

Baker, K. F. (1987). Evolving concepts of biological control of plant pathogens. *Annual Review of Phytopathology*, **25**, 67–85.

Baker, K. F. & Snyder, W. C. (ed.) (1965). *Ecology of soil-borne plant pathogens, prelude to biological control*. London: Murray.

Baker, K. F. & Cook, R. J. (1974). *Biological control of plant pathogens*. San Francisco: Freeman.

Baker, R. (1986). Biological control: an overview. *Canadian Journal of Plant Pathology*, **8**, 218–21.

Bohlool, B. B., & Schmidt, E. L. (1980). The immunofluorescence approach in microbial ecology. *Advances in Microbial Ecology*, **4**, 203–41.

Brill, W. J. (1985). Safety concerns and genetic engineering in agriculture. *Science*, **227**, 381–4.

Bruehl, G. W. (ed.) (1975). *Biology and control of soil-borne plant pathogens*. St Paul, Minnesota: American Phytopathological Society.

Campbell, R. (1986). The search for biological control agents: a pragmatic approach. *Biological Agriculture and Horticulture*, **3**, 317–27.

Cook, R. J. & Baker, K. F. (1983). *The nature and practice of biological control of plant pathogens*. St Paul, Minnesota: American Phytopathological Society.

Corke, A. T. K. & Rishbeth, J. (1981). Use of microorganisms to control plant diseases. In *Microbial control of pests and plant diseases*, ed. H. D. Burges, pp. 717–36. London: Academic Press.

Crespi, R. S. (1985). Microbiological inventions and the patent law – the international dimension. *Biotechnology and Genetic Engineering Reviews*, **3**, 1–37.

Deacon, J. W. (1983). *Microbial control of plant pests and diseases*. Aspects of Microbiology, 7. Wokingham: Van Nostrand Rheinhold.

Delp, C. J. (1977). Privately supported disease management activities. In *Plant disease*, vol. 1, ed. J. G. Horsfall & E. B. Cowling, pp. 381–92. New York: Academic Press.

Dhingra, O. D. & Sinclair, J. B. (1985). *Basic plant pathology methods*, pp. 245–58. Boca Raton. Florida: CRC Press.

Hartley, C. (1921). Damping off in forest nurseries. *USDA Bulletin*, **934**, 1–99.

Henry, A. W. (1931). The natural microflora of the soil in relation to the foot rot problem of wheat. *Canadian Journal of Research*, **4**, 69–77.

Lewis, C. J. (1977). The economics of pesticide research. In *Origins of pest, parasite, disease and weed problems*, ed. J. M. Cherrett & C. R. Sagar, pp. 237–45. Oxford: Blackwell Scientific Publications.

Lindemann, J. (1985). Genetic manipulation of microorganisms for biological control. In *Biological control on the phylloplane*, ed. C. E. Windels & S. E. Lindow, pp. 116–30. St Paul, Minnesota: American Phytopathological Society.

McKinney, H. H. (1929). Mosaic disease in the Canary Islands, West Africa and Gibraltar. *Journal of Agricultural Research* (Washington), **39**, 557–78.

Millard, W. A. & Taylor, C. B. (1927). Antagonism of microorganisms as the controlling factor in the inhibition of scab by green manuring. *Annals of Applied Biology*, **14**, 202–16.

Papavizas, G. C. (1985). Soilborne plant pathogens: new opportunities for control. British crop protection conference 1984. *Pests and diseases*, vol. 1, pp. 371–8. Croydon: BCPC Publications.

Parker, C. A., Rovira, A. D., Moore, K. J., Wong, P. T. W. & Kollmorgen, J. F. (ed.) (1985). *Ecology and management of soilborne plant pathogens*. St Paul, Minnesota: American Phytopathological Society.

Pimental, D., Glenister, C., Fast, S. & Gallahan, D. (1983). An environmental risk assessment of biological and cultural control in organic agriculture. In *Environmentally sound agriculture*, ed. W. Lockeretz, pp. 73–90. New York: Praeger Publishers.

Rishbeth, J. (1963). Stump protection against *Fomes annosus* III. Inoculation with *Peniophora gigantea*. *Annals of Applied Biology*, **52**, 63–77.

Ruffles, G. (1986). Patents and the biologist. *Biologist*, **33**, 5–10.

Sanford, G. B. (1926). Some factors affecting the pathogenicity of *Actinomyces scabies*. *Phytopathology*, **16**, 525–47.

Scher, F. M. & Castagno, J. R. (1986). Biocontrol: a view from industry. *Canadian Journal of Plant Pathology*, **8**, 222–4.

Schippers, B. & Gams, W. (ed.) (1979). *Soil-borne plant pathogens*. London: Academic Press.

Toussoun, T. A., Bega, R. V. & Nelson, P. E. (ed.) (1970). *Root diseases and soil-borne pathogens*. Berkeley: University of California Press.

Weller, D. M., Zhang, B-X. & Cook, R. J. (1985). Application of a rapid screening test for selection of bacteria suppressive to take-all of wheat. *Plant Disease*, **69**, 710–14.

Chapter 3

Austin, B., Dickinson, C. H. & Goodfellow, M. (1977). Antagonistic interactions of phylloplane bacteria with *Drechslera dictyoides* (Drechsler) Shoemaker. *Canadian Journal of Microbiology*, **23**, 710–15.

Bailey, J. H. & Deverall, B. J. (ed.) (1983). *The dynamics of host defence*. London: Academic Press.

Blakeman, J. P. (ed.) (1981). *Microbial ecology of the phylloplane*. London: Academic Press.

Blakeman, J. P. (1982). Phylloplane interactions. In *Phytopathogenic prokaryotes*, ed. M. S. Mount & G. H. Lacy, pp. 307–33. New York: Academic Press.

Burchill, R. T. & Cook, R. T. A. (1971). The interaction of urea and microorganisms in suppressing the development of perithecia of *Venturia inequalis* (Cke.) Wint. In *Ecology of leaf surface microorganisms*, ed. T. F. Preece & C. H. Dickinson, pp. 471–83. London: Academic Press.

Campbell, R. (1985). *Plant microbiology*. London: Arnold.

Cullen, D. & Andrews, J. H. (1984). Epiphytic microbes as biological control agents. In *Plant-microbe interactions*, ed. T. Kosuge & E. W. Nester, pp. 381–99. New York: MacMillan.

Dean, R. A. & Kuć, J. (1986). Induced systemic protection in cucumber: time of production and movement of the signal. *Phytopathology*, **76**, 966–70.

Dickinson, C. H. & Preece, T. F. (1976). *Microbiology of aerial plant surfaces*. London: Academic Press.

Fokkema, N. J. (1973). The role of saprophytic fungi in antagonism against *Drechslera sorokiniana* (*Helminthosporium sativum*) on agar plates and on rye leaves with pollen. *Physiological Plant Pathology*, **3**, 195–205.

Fokkema, N. J., Houter, J. G. den, Kosterman, Y. J. C. & Nelis, A. L. (1979). Manipulation of yeasts on field-grown wheat leaves and their antagonistic effect on *Cochliobolus sativus* and *Septoria nodorum*. *Transactions of the British Mycological Society*, **72**, 19–29.

Fokkema, N. J. & Nooij, M. P. de (1981). The effect of fungicides on the microbial balance in the phyllosphere. *European and Mediterranean Plant Protection Organisation Bulletin*, **11**, 303–10.

Fokkema, N. J. & Heuval, J. van den. (1987).*Microbiology of the phyllosphere.* Cambridge: Cambridge University Press.

Fokkema, N. J., Laar, J. J. van de, Nelis-Blomberg, A. L. & Schippers, B. (1975). The buffering capacity of the natural mycoflora of rye leaves to infection by *Cochliobolus sativus*, and its susceptibility to benomyl. *Netherlands Journal of Plant Pathology*, **81**, 176–86.

Johnston, A. & Booth, C. (ed.) (1983). *Plant pathologist's pocketbook.* Slough: Commonwealth Agricultural Bureaux.

Kranz, J. (1981). Hyperparasitism of biotrophic fungi. In *Microbial ecology of the phylloplane*, ed. J. P. Blakeman, pp. 327–52. London: Academic Press.

Kuć, J. (1981). Multiple mechanisms, reaction rates and induced resistance in plants. In *Plant disease control*, ed. R. C. Staples & G. H. Toenniessen, pp. 259–84. New York: Wiley.

Lindow, S. E. (1983). The role of bacterial ice nucleation in frost injury to plants. *Annual Review of Phytopathology*, **21**, 363–84.

Lindow, S. E. (1985). Integrated control and the role of antibiosis in biological control of fireblight and frost injury. In *Biological control on the phylloplane*, ed. C. E. Windels & S. E. Lindow, pp. 83–115. St Paul, Minnesota: American Phytopathological Society.

Lockwood, J. L. & Filonow, A. B. (1981). Responses of fungi to nutrient limiting conditions and to inhibitory substances in natural habitats. *Advances in Microbial Ecology*, **5**, 1–61.

Mulinge, S. K. & Griffiths, E. (1974). Effects of fungicides on leaf rust, berry disease, foliation and yield of coffee. *Transactions of the British Mycological Society*, **62**, 495–507.

Newhook, F. J. (1951). Microbiological control of *Botrytis cinerea* Pers. II Antagonism by fungi and actinomycetes. *Annals of Applied Biology*, **38**, 185–202.

Preece, T. F. & Dickinson, C. H. (1971). *Ecology of leaf surface microorganisms.* London: Academic Press.

Ruinen, J. (1961). The phyllosphere 1. An ecologically neglected milieu. *Plant and Soil*, **15**, 81–109.

Scherff, R. H. (1973). Control of bacterial blight of soybean by *Bdellovibrio bacteriovorus*. *Phytopathology*, **63**, 400–02.

Sequeira, L. (1983). Mechanisms of induced resistance in plants. *Annual Review of Microbiology*, **37**, 51–79.

Windels, C. E. & Lindow, S. E. (ed.) (1985). *Biological control on the phylloplane.* St Paul, Minnesota: American Phytopathological Society.

Wood, R. K. S. (1951). The control of diseases of lettuce by use of antagonistic organisms. 1. Control of *Botrytis cinerea* Pers. *Annals of Applied Biology*, **38**, 203–16.

Chapter 4

Anagnostakis, S. (1982). Biological control of chestnut blight. *Science*, **215**, 466–71.

Cook, R. J. & Baker, J. F. (1983). *The nature and practice of biological control of plant pathogens.* St Paul, Minnesota: American Phytopathological Society.

Cooke, R. C. & Rayner, A. D. M. (1984). *Ecology of saprotrophic fungi*. London: Longmans.

Moore, L. W. & Cooksey, D. A. (1981). Biology of *Agrobacterium tumefaciens*. In *Biology of Rhizobiaceae*. ed. K. L. Giles & A. G. Atherley, pp. 15–46. *International Review of Cytology*, Supplement 13.

Rayner, A. D. M. & Boddy, L. (1986). Population structure and the infection biology of wood decay fungi in living trees. *Advances in Plant Pathology*, **5**, 119–60.

Rishbeth, J. (1963). Stump protection against *Fomes annosus*. III Inoculation with *Peniophora gigantea*. *Annals of Applied Biology*, **52**, 63–77.

Thomson, J. A. (1987). The use of agrocin-producing bacteria in the biological control of crown gall. In *Innovative approaches to plant disease control*, ed. I. Chet., pp. 213–28. New York: John Wiley.

van Alfen, N. K. (1982). Biology and potential for disease control of hypovirulence of *Endothia parasitica*. *Annual Review of Phytopathology*, **20**, 349–62.

Chapter 5

Alabouvette, C., Rouxel, F. & Louvet, J. (1979). Characteristics of *Fusarium* wilt-suppressive soils and proposals for their utilization in biological control. In *Soil-borne plant pathogens*, ed. B. Schippers & W. Gams, pp. 165–82. London: Academic Press.

Alabouvette, C., Couteaudier, Y. & Louvet, J. (1985). Soils suppressive to *Fusarium* wilt: mechanisms and management of suppressiveness. In *Ecology and management of soilborne plant pathogens*, ed. C. A. Parker, *et al.*, pp. 101–6. St Paul, Minnesota: American Phytopathological Society.

Asher, M. J. C. & Shipton, P. J. (ed.) (1981). *Biology and control of take-all*. London: Academic Press.

Ayers, W. A. & Adams, P. B. (1981). Mycoparasitisms and its application to biological control of plant diseases. In *Biological control in crop production*, ed. G. C. Papavizas, pp. 91–105. *Beltesville Symposium in Agricultural Research* 5. Granada: Allenheld, Osmum Publishing.

Bagyaraj, D. (1984). Biological interactions with VA mycorrhiza. In *VA mycorrhiza*. ed. C. L. P. Powell & D. Bagyaraj, pp. 131–53. Boca Raton, Florida: CRC Press.

Bruehl, G. W. (ed.) (1987). *Soilborne plant pathogens*. New York: Macmillan Publishing.

Campbell, R. (1983). *Microbial ecology*. Oxford: Blackwell.

Campbell, R. (1985). *Plant microbiology*. London: Edward Arnold.

Chet, I. (ed.) (1987). *Innovative approaches to plant disease control*. New York: John Wiley.

Chet, I. & Henis, Y. (1985). *Trichoderma* as a biocontrol agent against soilborne root pathogens. In *Ecology and management of soilborne plant pathogens*, ed. C. A. Parker, *et al.*, pp. 110–12. St Paul, Minnesota: American Phytopathological Society.

Cook, R. J. (1986). Plant health and the sustainability of agriculture, with special reference to disease control by beneficial organisms. *Biological Agriculture and Horticulture*, **3**, 211–32.

Cook, R. J. & Baker, K. F. (1983). *The nature and practice of biological control of plant pathogens*. St Paul, Minnesota: American Phytopathological Society.

Curl, E. A. & Truelove, B. (1985). *The rhizosphere*. Berlin: Springer-Verlag.

Garrett, S. D. (1970). *Pathogenic root infecting fungi*. Cambridge: Cambridge University Press.

Gerlagh, M. (1968). Introduction of *Ophiobolus graminis* into new polders and its decline. *Netherlands Journal of Plant Pathology*, **74**, Suppl. 2, 97 pp.

Hoitink, H. A. J. & Fahy, P. C. (1986). Basis for control of soilborne plant pathogens with composts. *Annual Review of Phytopathology*, **24**, 93–114.

Hornby, D. (1979). Take-all decline: a theorist's paradise. In *Soil-borne plant pathogens*, ed. B. Schippers & W. Gams, pp. 133–56. London: Academic Press.

Hornby, D. (1985). Soil nutrients and take-all. *Outlook on Agriculture*, **14**, 12–128.

Lynch, J. M. (1987). Biological control within microbial communities of the rhizosphere. In *Ecology of microbial communities*, ed. M. Fletcher, T. R. G. Gray & J. G. Jones, pp. 55–82. Society for General Microbiology, Symposium 41. Cambridge: Cambridge University Press.

Malajczuk, N. (1979). Biological suppression of *Phytophthora cinnamomi* in eucalypts and avocado in Australia. In *Soil-borne plant pathogens*, ed. B. Schippers & W. Gams, pp. 635–52. London: Academic Press.

Nedwell, D. B. & Gray, T. R. G. (1987). Soils and sediments as matrices for microbial growth. *In* Fletcher *et al.* see Lynch above. pp. 21–54.

Old, K. M. & Chakraborty, S. (1986). Mycophagous soil amoebae; their biology and significance in the ecology of soil-borne plant pathogens. *Progress in Protistology*, **1**, 163–94.

Palti, J. (1981). *Cultural practices and infectious crop diseases*. Berlin: Springer-Verlag.

Parker, C. A., Rovira, A. D., Moore, K. J., Wong, P. T. W. & Kollmorgan, J. F. (ed.) (1985). *Ecology and management of soilborne plant pathogens*. St Paul, Minnesota: American Phytopathological Society.

Rovira, A. D. & Wildermuth, G. B. (1981). The nature and mechanisms of suppression. *In* Asher, M. J. C. & Shipton, P. J. (ed.) *loc. cit.* pp. 385–415.

Schippers, B. & Gams, W. (ed.) (1979). *Soil-borne plant pathogens*. London: Academic Press.

Schippers, B., Bakker, A. W. & Bakker, P. H. A. M. (1987). Interaction of deleterious and beneficial rhizosphere microorganisms and their effect on the cropping of potatoes. *Annual Review of Phytopathology*, **25**, 339–58.

Schneider, R. W. (ed.) (1982). *Suppressive soils and plant disease*. St Paul, Minnesota: American Phytopathological Society.

Swinburne, T. R. (ed.) (1986). *Iron, siderophores and plant diseases*. New York: Plenum Press.

Weller, D. M. (1985). Application of fluorescent pseudomonads to control root diseases. In *Ecology and management of soilborne plant pathogens*, ed. C. A. Parker, *et al.*, pp. 137–40. St Paul, Minnesota: American Phytopathological Society.

Weller, D. M. & Cook, R. J. (1986). Suppression of root diseases of wheat by fluorescent pseudomonads and mechanisms of action. In *Iron, siderophores and plant diseases*, ed. T. R. Swinburne, pp. 99–107. New York: Plenum Press.

Chapter 6

Beer, S. V., Rundle, J. R. & Norielli, J. L. (1984). Recent progress in the development of biological control for fire blight. *Acta Horticulturae*, **151**, 195–201.

Byrde, R. J. W. & Willetts, H. J. (1977). *The brown rot fungi of fruit: their biology and control*. Oxford: Pergammon Press.

Dennis, C. (ed.) (1983). *Post-harvest pathology of fruits and vegetables*. London: Academic Press.

Lindow, S. E. (1985). Integrated control and the role of antibiosis in biological control of fire blight and frost injury. In *Biological control on the phylloplane*, ed. C. E. Windels & S. E. Lindow, pp. 83–115. St Paul, Minnesota: American Phytopathological Society.

Purchase, I. F. H. (1974). *Mycotoxins*. Amsterdam: Elsevier.

Swinburne, T. R. (1986). Stimulation of disease development by siderophores and inhibition by chelated iron. In *Iron, siderophores and plant disease*. ed. T. R. Swinburne, pp. 217–26. NATO ASI series vol. 117. New York: Plenum Press.

Chapter 7

Castanho, B. & Butler, E. E. (1978). *Rhizoctonia* decline: studies on hypovirulence and potential use in biological control. *Phytopathology*, **68**, 1511–14.

Cook, R. J. & Baker, J. F. (1983). *The nature and practice of biological control of plant pathogens*. St Paul, Minnesota: American Phytopathological Society.

Gerlagh, M. (1968). Introduction of *Ophiobolus graminis* into new polders and its decline. *Netherlands Journal of Plant Pathology*, **74**, suppl. 2, 97 pp.

Kommedahl, T. & Windels, C. E. (1981). Introduction of microbial antagonists to specific courts of infection: seeds, seedlings and wounds. In *Biological control in crop production*, ed. G. C. Papavizas, pp. 227–48. *Beltesville Symposia in Agricultural Research 5*. Granada: Allenheld, Osmum Publishing.

Papavizas, G. C. (1985). *Trichoderma* and *Gliocladium*: biology, ecology and potential for biological control. *Annual Review of Phytopathology*, **23**, 23–54.

Schneider, R.N. (ed.) (1982). *Suppressive soils and plant disease*. St Paul, Minnesota: American Phytopathological Society.

Expanded index of pathogens

This index gives information on the plant pathogens mentioned in the text, including the genus name, authority, species and subdivisions of the species (variety, *forma specialis* etc.), the disease caused, its importance and distribution. Main references are given as follows: Commonwealth Mycological Institute, Kew, England, *Descriptions of Pathogenic Fungi and Bacteria* (CMI —); Commonwealth Agricultural Bureau, *Mycological Papers*, Kew (CAB MP —); Centraalbureau voor Schimmelcultures, Baarn, *Studies in Mycology* (CVS SM —); Starr, M. P. *et al.* (eds.) 1981, *The Prokaryotes*. Berlin: Springer-Verlag; Ainsworth, G. C. (ed.) 1971, *Dictionary of Fungi*. CMI, Kew has been used extensively and there are individual references in some cases.

Agrobacterium Conn. Bacterium, Rhizobiaceae. *A. tumefaciens* (CMI 42; Starr p. 842) and *A. rhizogenes* cause crown gall and hairy root respectively. Important worldwide, Ti plasmid used as a vector in genetic engineering. Control by hygiene and the use of avirulent strains (K84) of *A. radiobacter*. pp. 34, 106–110, 133.

Alternaria Nees ex Wallr. Fungus, deuteromycete. *A. brassicae* on cabbages (CMI 162), *A. brassicicola* on cabbages (CMI 163), *A. citri* on citrus (CMI 242), *A. crassa* (CMI 243) and *A. cucumerina* (CMI 244) and many more on a wide variety of hosts worldwide. Cause leaf spots, fruit and flower rots, usually seed borne. Control by a variety of fungicides (not benomyl) and some host resistance. See also CAB MP 20. pp. 71, 74, 78.

Armillaria (Fr.) Stande. Fungus, basidiomycete, Agaricales. *A. mellea*, honey fungus (CMI 321) causing rot of roots and stems of trees and shrubs, spreads by rhizomorphs. Control by soil sterilization in special cases, otherwise none. pp. 6, 18, 38, 157–8.

Aspergillus Mich. ex Fr. Fungus, deuteromycete. (CAB MP 2; Raper, K. B. & Fennell, D. I. (1965). *The genus* Aspergillus, Baltimore: Williams and Wilkins; Domsch, K. H., Gams, W. & Anderson, T-H. (1980). *Compendium of soil fungi*. pp. 76–124. London: Academic Press). Mostly saprotrophs but *A. flavus* and others produce mycotoxins causing spoilage of fruits and seeds. Worldwide distribution. Control by storage conditions and many fungicides. p. 167.

Botrytis Mich. ex Fr. Fungus, mostly deuteromycetes (Ellis, M. B. (1971). *Dematiaceous hyphomycetes* Kew, CMI). *B. cinerea* (Domsch, K. H., Gams,

W. & Anderson, T-H. (1980). *Compendium of soil fungi*. pp. 146–55. London: Academic Press) perfect state *Sclerotinia fuckeliana* (CMI 431). *B. fabae* (CMI 432; CAB MP 62), also *B. narcissicola*. Saprotrophs and pathogens of a wide variety of hosts worldwide. Cause fruit decay, leaf spots, bulb rots, etc. Control by many fungicides, though have developed serious resistance to some. pp. 5, 71, 78–9, 82, 99, 140, 149, 164–7.

Cephalosporium Corda. Fungus, deuteromycete. *C. gramineum* (CMI 501) perfect state *Hymenula cerealis*, ascomycete, causes leaf stripe on wheat, barley, oats and other cereals. Widely distributed but only limited damage. pp. 24, 38, 158.

Ceratocystis Ellis & Habst. Fungus, ascomycete. *C. ulmi* (CMI 361; Phillips, D. H. & Burdekin, D. A. (1982). *Diseases of forest and ornamental trees*. p. 262. Basingstoke: Macmillan) causes Dutch elm disease, introduced into many countries and kills most elms. Strains vary in virulence, hypovirulent strains are reported. Very important, removes all elms of useful size from large areas of the temperate world. Some systemic fungicides effective in delaying disease, but not economic, control of the beetle vector helps. pp. 104–5.

Chondrostereum Pouzar. Fungus, basidiomycete (= *Stereum*). *C. purpureum* causes silverleaf of fruit trees and other broad leaved trees (Phillips, D. H. & Burdekin, D. A. (1982). *Diseases of forest and ornamental trees*. Basingstoke: Macmillan.) Worldwide distribution. Control by some fungicides and by *Trichoderma*. pp. 96, 106.

Cladosporium Link ex Fr. Fungus, deuteromycete (Ellis, M. B. (1971). p. 308. *Dematiaceous hyphomycetes* Kew: CMI.) *C. herbarum* and *C. cladosporioides* are very common saprotrophs. *C. fulvus* causes tomato leaf mould and *C. cucumerinum* causes cucumber gummosis. Control by fungicides. pp. 10, 74.

Claviceps Tul. Fungus, ascomycete. *C. purpurea* causes ergot of cereals (James, D. G. & Clifford, B. C. (1978). *Cereal diseases, their pathology and control*. p. 169. UK: BASF Ltd.) Controlled by seed hygiene. p. 140.

Cochliobolus Drechs. Fungus, ascomycete. *C. sativus* (CMI 701) causes foot and root rot of temperate cereals. *C. miyabeanus* (CMI 302) causes seedling blight of rice. Controlled by fungicide seed treatment, also known to be parasitized by amoebae. pp. 72–4, 83–4, 149.

Colletotrichum Corda. Fungus, deuteromycete. *C. lindemuthianum* (CMI 316) causes anthracnose of beans, widely distributed and important, controlled by seed hygiene, host resistance and induced resistance. *C. lagenarium* causes anthracnose of cucumbers. *C. musae* (CMI 222) causes anthracnose in bananas for which there is no good control. *C. cucumerinum* and *C. coffeanum* also important. Have particularly been used for studies on induced resistance and cross-protection generally. pp. 32, 78, 84–7, 164, 166.

Corynebacterium Lehmann & Neumann. Bacterium, Corynebacteriaceae (Starr *et al.*, p. 1827). Produce various toxins and cause wilts in general. *C. michiganense* (Starr *et al.*, p. 1879; Billing, E. (1987). *Bacteria as plant pathogens* Aspects of Microbiology 14. Wokingham: Van Nostrand Rheinhold; CMI 19) causes canker of tomato and pepper. p. 115.

Drechslera Ito. Fungus, deuteromycete (Ellis, M. B. (1971). *Dematiaceous hyphomycetes* p. 403. Kew: CMI). *D. dictyoides* (perfect state *Pyrenophora dictyoides*) causes net blotch of some grasses and cereals. pp. 79–81.

Endothia Fr. Fungus, deuteromycete. *E. parasitica* causes chestnut blight, a canker of stems of many species of chestnut worldwide, now introduced into many temperate countries and kills all chestnut trees of useful size. Isolates vary in virulence, hypovirulent strains known and used in control. pp. 102–4, 133.

Erwinia Winslow *et al.* Bacterium, Enterobacteriaceae (Starr *et al.*, p. 1260). *E. amylovora* (CMI 44) causes fire blight of rosaceous trees and shrubs, especially important on fruit trees such as pear, worldwide. Control by hygiene, the use of antibiotic sprays in USA, and biological means. *E. carotovora* var. *atroseptica* (CMI 551) causes black leg of potato controlled by seed piece hygiene. *E. carotovora* var. *carotovora* (CMI 552) causes various soft rots of fruits and vegetables controlled by hygiene in packing and handling. *E. herbicola* (CMI 232) is a very common saprotroph used in biological control of *E. amylovora*. pp. 90–2, 106, 162–4.

Erysiphe Hedw. ex Fr. Fungus, ascomycete (Spencer, D. M. (1978). *The powdery mildews* London: Academic Press). *E. graminis* (CMI 153) causes mildew on cereals and grasses worldwide, very important cause of yield loss. There are specific mildew fungicides as seed treatment or foliar spray and host resistance. pp. 9, 69, 70.

Eutypa Tul. Fungus, ascomycete. *E. armeniacae* (CMI 436) causes gummosis (the production of gums and resin), cankers or die-back of apricot trees worldwide. Lives as a saprotroph on many rosaceous hosts and on vines, walnut etc. Control by the treatment, biological and chemical, of pruning wounds. pp. 96–8.

Fistulina Bull. ex Fr. Fungus, basidiomycete. *F. hepatica* causes a timber rot, especially of oak trees. p. 17.

Fusarium Link ex Fr. Fungus. deuteromycete (Booth, C. (1971). *The genus Fusarium*, Kew: CMI; Booth, C. (1977). Fusarium Kew: CMI). A very common genus of pathogens causing wilts and rots, and also many species of saprotrophs. Many species of pathogens have numerous *formae speciales*. Worldwide distribution and often very important. *F. culmorum* (CMI 26) occurs on a wide range of host families (cereals, brassicas, composites, conifers, legumes, Rosaceae, vines, etc.) with seedling blights, foot rots and root rots amongst other diseases; especially important on cereals. *F. solani* (CMI 29) causes infections of wounds and weakened plants, also causes root and stem rots, very wide host range but especially important on peas after infection with *F. oxysporum* f. sp. *pisi*. *F. roseum* is a group of species (including *avenaceum, culmorum, graminearum*, etc.) but still used as a species in some of the older literature. *F. nivale* (perfect state = *Micronectria nivalis*) causes snow mould, pre-emergence blight and root rot of wheat and rye, serious on winter cereals, especially under snow and at low temperatures. *F. oxysporum* (CMI 211) is the most important pathogenic species causing wilt of a variety of crops: f. sp. *cubense* (CMI 214) of bananas is very important and is controlled by a number of biological methods; f. sp. *cucumerinum* occurs on cucurbits other than melon; f. sp. *lini* on flax; f. sp. *lycopersici* (CMI 217) causes vascular wilt of tomatoes, though there are resistant cultivars; f. sp. *melonis* (CMI 218) on melons where it causes wilt; f. sp. *pisi* causes wilt of peas. pp. 7, 13, 18–9, 23, 26, 32, 95, 106, 115, 119–24, 128, 130–1, 133, 138, 144–8, 157, 169 *et seq.*

Gaeumannomyces Arx & Olivier. Fungus, ascomycete (Asher, M. J. C. & Shipton, P. J. (1981) *Biology and control of take-all*. London: Academic Press).

G. graminis var. *tritici* (CMI 383) causes take-all of wheat and barley, var. *avenae* (CMI 382) on oats and var. *graminis* on grasses, though occurs avirulent on wheat. Important worldwide in temperate wheat growing areas. The asexual states are *Phialophora* amongst the many species and subsections of which there is much taxonomic confusion (Walker, in Asher & Shipton, see above): *Phialophora* is also known as an imperfect form-genus (i.e. no known perfect states, *Gaeumannomyces* or otherwise). No cultivar resistance or chemical control, much worked on for biological control. pp. 6, 13, 15, 18, 21, 25, 39, 115, 117–9, 124–7, 131–2, 134–6, 148–9, 157.

Gliocladium Corda. Fungus, deuteromycete. *G. roseum* is a saprotroph or occasionally a pathogen. May colonize seeds germinating under cold, wet conditions and increase oxygen stress. Also a mycoparasite and general antagonist, see antagonist index. p. 18.

Helminthosporium Link. Fungus, deuteromycete (CAB MP 158). Many common saprotrophs, but *H. solani* (CMI 167) causes silver scurf of potatoes. Many of the species, including some pathogens, are now considered to belong to the genus *Drechslera* (Ellis, M. B. (1971). *Dematiaceous hyphomycetes*, Kew: CMI). pp. 71, 74.

Hemileia Berk. & Br. Fungus, basidiomycete, Uredinales. *H. vastatrix* (CMI 1) causes coffee rust in Asia, Africa and now S. America. Very serious. Some control by fungicides. pp. 69, 75.

Heterobasidion Bref. (= *Fomes* (Fr.) Cooke). Fungus, basidiomycete. *H. annosum* (CMI 192; Phillips, D. H. & Burdekin, D. A. (1982). *Diseases of forest and ornamental trees*. Basingstoke: Macmillan Press) causes a serious butt rot of conifers in temperate forests. Control by biological treatment of freshly cut stumps or by using chemicals that favour saprotrophic colonization. pp. 43, 99–101.

Melampsora Cast. Fungus, basidiomycete, Uredinales. *M. lini* (CMI 51; Wilson, M. & Henderson, D. M. (1966) *British rust fungi*. Cambridge: Cambridge University Press) causes flax rust worldwide, an important disease. The fungus is also much used in genetic research on virulence and host–pathogen recognition etc. p. 69.

Meria Vuill. Fungus, deuteromycete. *M. laricis* is a pathogen of larch needles. Controlled by sanitation and a variety of fungicides. p. 79.

Monilinia Honey. Fungus, deuteromycete, perfect stage *Sclerotinia* (Byrde, R. J. W. & Willetts, H. J. (1977). *The brown rot fungi of fruit*. Oxford: Pergamon Press). *M. fructicola* (= *S. fructicola*, CMI 616) and *M. fructigena* (= *S. fructigena*, CMI 617) both cause brown rots of fruits, especially apples, plums, apricots, peach, pear, etc. Control by hygiene and fungicides. pp. 164–5, 168.

Mucor Mich. ex St.-Am. Fungus, zygomycete (CVS SM 10, 12, 17; CMI 527, 528; Domsch, K. H., Gams, W. & Anderson, T-H. (1980). *Compendium of soil fungi* pp. 461–80. London: Academic Press). Mostly saprotrophs but some species can cause rots of fruits, especially after damage in picking or handling. pp. 165–6.

Nectria Fr. Fungus, ascomycete (CAB MP 73, 150). *N. galligena* causes canker of fruit trees, very common, controlled by fungicides. Also *N. cinnabarina* which causes coral spot and many saprotrophic species. pp. 97–8.

Penicillium Link ex Fr. Fungus, deuteromycete (CVS SM 23; Pitt, J. I. (1979)

The genus Penicillium. London: Academic Press; CMI 96–9). Perfect states are *Eupenicillium* and *Talaromyces*. Very common saprotrophs, worldwide. Also cause fruit rots. Used also as an antagonist, see antagonist index. pp. 7, 164.

Phialophora Medlar. Fungus, deuteromycete. See under *Gaeumannomyces*.

Phymatotrichum Bon. Fungus, deuteromycete. *P. omnivorum* (= *Phymatotrichopsis omnivora* in Domsch, K. H., Gams, W. & Anderson, T-H. (1980). *Compendium of soil fungi*. London: Academic Press). Causes a root rot of cotton and very many other plants, hence the species name. Important on cotton in southern USA. pp. 13, 37, 119, 128.

Phytophthora de Bary. Fungus, oomycete (CAB MP 92, 12, 143). Very important pathogens worldwide. *P. infestans* (CMI 838) causes potato blight, controlled by fungicides and complex patterns of host resistance to races of the pathogen of different virulence. Still causes losses, caused the Irish potato famine and hence is responsible for a not inconsiderable part of the population of the USA because of mass emigration. *P. cinnamomi* (CMI 113) causes root rot and death of many trees and shrubs; recently important in Australia for the destruction of *Eucalyptus* forests. No chemical control, biological methods being developed. pp. 11, 13, 24, 29, 70, 115, 127, 130–1, 149, 169 *et seq.*, 174.

Plasmodiophora Woron. Fungus, Plasmodiophoromycetes. *P. brassicae* (CMI 621) causes club root of all brassicas (with the exception of black mustard) very little host resistance but the disease rarely fatal. Important worldwide as control is difficult; fungicide root dips at planting help, spores persist in the soil for 10's of years as resistant cysts, so strict hygiene necessary. p. 35.

Pseudocercosporella Deighton. Fungus, deuteromycete (CAB MP 133). *P. herpotrichoides* (CMI 386) causes eyespot of cereals and many wild grasses, widespread and important in temperate areas of the world. Control by systemic fungicides but resistance to them has developed. pp. 35, 38, 157.

Pseudomonas Migula. Bacterium, pseudomonads (Starr *et al.*, pp. 656–741, especially 701–18; Billing, E. (1987). *Bacteria as plant pathogens*. UK: Van Nostrand Rheinhold). *P. glycinae* attacks soybean. *P. syringae* (CMI 46) causes rots of stone fruits. *P. solanacearum* (CMI 15) causes rots and wilts of many plants in warm temperate, tropical and sub-tropical regions. *P. tolaasii* (CMI 894) causes brown blotch disease of mushrooms. *P. fluorescens* is mostly saprotrophic (see antagonist index) and there are many other species and subdivisions thereof. The most important genus of bacteria attacking plants. Worldwide. Few effective control measures. pp. 33, 49, 88, 90, 115.

Pseudoperonospora Rostortsev. Fungus, oomycete (Spencer, D. M. (ed.) (1981). *The downy mildews*. London: Academic Press). *P. humuli* (CMI 769) causes downy mildew of hops, widespread and important, control by host resistance. *P. cubense* (CMI 457) causes decay and mildew of cucurbits worldwide, controlled by fungicides, host resistance and cultural methods. pp. 13, 69.

Puccinia Pers. Fungus, basidiomycete, Uredinales (Jones, D. G. & Clifford, B. C. (1978). *Cereal diseases, their pathology and control*. UK: BASF; Wilson, M. & Henderson, D. M. (1966). *British rust fungi*. Cambridge: Cambridge University Press). *P. graminis* causes black stem rust of wheat and other cereals worldwide, can be serious but controlled by fungicides, plant resistance and the removal of the alternative host, barberry. *P. striiformis* (CMI 291) yellow rust of wheat and barley, important especially in cool moist conditions worldwide, controlled by fungicides and plant resistance. pp. 29, 69, 90.

Pythium Pringsheim. Fungus, oomycete (CAB MP 109, 110; CVS SM 21; Domsch, K. H., Gams, W. & Anderson, T-H. (1980). *Compendium of soil fungi*, pp. 678–97. London: Academic Press). A widespread and important genus of soil-borne pathogens, especially serious in wet soils, causing pre- and post-emergence damping-off in seedlings. *P. ultimum* especially in temperate conditions with a very wide host range. pp. 5, 9, 11, 13, 23–4, 42, 115, 123, 128, 130–1, 138, 148, 169 *et seq.*, 173–6, 179.

Rhizoctonia DC ex Fr. Fungus, deuteromycete, mycelia sterilia. *R. solani* (perfect state *Thanatephora cucumeris*, CMI 406) causes seed decay and damping-off in many hosts and later various stem decays (e.g. *R. cerealis*, sharp eyespot on cereals, which was until recently part of *R. solani*). Worldwide distribution and very common in soil. Controlled by fumigation in nurseries, many antagonists known. pp. 5, 12–3, 18, 25, 36–7, 71, 74, 82, 113, 124, 128, 131, 138–9, 140, 142–3, 148–9, 157–8, 169, 172 *et seq.*

Rhizopus Ehrenb. ex Corda. Fungus, zygomycete (CVS SM 25; CMI 525, 526). Many species, mostly saprotrophs, but some attack stored fruits and vegetables. Very common worldwide. pp. 164–5.

Sclerotinia Fuckel. Fungus, ascomycete (CAB MP 62, p. 144). *S. minor* causes rots, especially of lettuce. *S. sclerotiorum* (CMI 513) causes rots of many vegetables with a very wide host range and worldwide. Controlled by fumigation in horticulture, and cultural treatments, though the sclerotia survive for long periods. pp. 130, 140–1, 171.

Sclerotium Tode ex Fr. Fungus, deuteromycete mostly. *S. rolfsii* (perfect state *Corticium rolfsii* [basidiomycete] CMI 410) causes root and stem rots, wilts etc. in a wide range of hosts. Controlled by soil fumigation and rotation in horticulture. *S. cepivorum* (CMI 512), white rot of onions, serious, persistent though fungicides give some control. *S. trifoliorum* especially on clover. pp. 13, 18, 25, 37, 138–40, 144, 173, 175.

Septoria Sacc. Fungus, deuteromycetes mostly. *S. nodorum* (= *Leptosphearia nodorum* [ascomycete] CMI 86) causes glume blotch of wheat and barley (Jones, D. G. & Clifford, B. C. (1978). *Cereal diseases, their pathology and control*. UK: BASF). Important on winter wheat worldwide, but especially in Europe. Controlled by fungicides. pp. 72, 83.

Sphearotheca Lev. Fungus, ascomycete (Spencer, D. M. (ed.) 1978. *The powdery mildews*. London: Academic Press). *S. fuliginea* causes powdery mildew on cucurbits worldwide. Controlled by host resistance and fungicides. p. 69.

Streptomyces Waksman & Henrici. Bacterium, actinomycete. *S. scabies* (Starr *et al.*, pp. 2028–90 especially p. 2039) causes potato scab, mostly in alkaline soils. Controlled by the addition of organic matter. Common worldwide and causes yield loss and severe loss of quality, but not lethal. pp. 37, 41–2, 119, 127.

Taphrina Fr. Fungus, ascomycete (CMI 711) causes leaf curl in peaches, almonds, cherry, etc. Worldwide, common and serious in some orchards because of the premature defoliation caused. Control by fungicides. p. 9.

Thielaviopsis Went. Fungus, deuteromycete (CVS SM 8; CAB MP 83). *T. basicola* (CMI 170) causes black root rot of many plants, especially studied in tobacco. Worldwide distribution, control by plant resistance. pp. 13, 38, 131–2, 149.

Venturia de Not. Fungus, ascomycete. *V. inequalis* (CMI 401) causes scab of apples, on fruit leaves and also stems. Common worldwide. Control by fungicides. pp. 70, 76, 84–5, 98.

Verticillium Nees ex Wallr. Fungus, deuteromycete. Widespread, but mostly temperate, common and important pathogens causing wilts. *V. albo-atrum* (CMI 255) on many plant hosts, control by host resistance. *V. dahliae* (CMI 256) causes wilts of many host species and again control by resistant cultivars. pp. 13, 115, 133–5, 138, 149.

Expanded index of antagonists

This index lists all the antagonists named in the text. Abbreviations are as for the expanded index of pathogens (p. 200) and individual references are given where appropriate. In addition Cook, R. J. and Baker, K. F. 1983 (*The nature and practice of biological control of plant pathogens*. St Paul, Minnesota, American Phytopathological Society) has an extended list of antagonists.

Acanthamoeba Volkonsky. Protozoa, amoeba (Old, K. M. & Chakraborty, S. (1986). *Progress in Protistology* 1, 163–94; Page, F. C. (1977). *An illustrated guide to freshwater and soil amoebae*. Special Publication 34. UK: Freshwater Biological Association). This genus parasitizes fungi. Common and widespread in soil. Importance in biological control uncertain but may reduce dormant inoculum. Not commercial. p. 24.

Actinomyces Hartz. Bacterium (Starr *et al.*, 1981). Soil organism, not important. p. 42. See *Streptomyces*.

Agrobacterium Conn. Bacterium (Starr, *et al.*, 1981). *A. tumefaciens* is a pathogen, but strain 84 (= *A. radiobacter*) is avirulent and produces a bacteriocin (agrocin 84). Used to control crown gall. Commercially available in various formulations for root and cutting dips. One of the few examples of a biocontrol agent whose mechanism of action, genetics, etc. are quite well understood. pp. 34, 106 *et seq.*

Alternaria Nees ex Wallr. Fungus, deuteromycete. Genus contains both pathogens and saprotrophs. Various species of the latter have been used experimentally, especially on leaves. pp. 84–5.

Ampelomyces Ces. Fungus, deuteromycete. *A. quisqualis* (perfect state *Cicinnobolus cesatiî*, an ascomycete) is a mycoparasite, especially on powdery mildews (Kranz, J. (1981). In *Microbial ecology of the phylloplane*, ed. J. P. Blakeman. London: Academic Press). Mycoparasitism usually occurs late in the infection by the mildew, so reduces inoculum rather than controls the present disease. p. 89.

Arachnula Cienk. Protozoa, amoeba (Old, K. M. & Darbyshire, J. F. (1980). *Protistologia* 16, 277–87; Old, K. M. & Chakraborty, S. (1986). *Progress in Protistology* 1, 163–94). Parasite of fungal spores (e.g. *Cochliobolus sativus*). Common and widespread, but importance unknown. pp. 24, 149.

Aspergillus Mich. ex Fr. Fungus, deuteromycete (see pathogen index for references). Common soil saprotroph, occurs in lists of isolates and potential antagonists, but not important. pp. 123, 167.

Aureobasidium Viala & Boyer. Fungus, denteromycete yeast (CVS SM 15). Common leaf surface saprotroph, used experimentally not commercially. pp. 68, 72–3, 77, 81.

Bacillus Cohn. Bacterium (Starr *et al.*, 1981). *B. cereus* (including var. *mycoides*), *B. subtilis* and *B. pumilus* occur worldwide in many soils and have been proposed as biocontrol agents against *Sclerotinia, Gaeumannomyces, Fusarium*, etc. Also used to protect tree wounds. Have reached field trials stage in many cases. Produce antibiotics, easy to grow and isolate and have a long shelf-life. Much studied though not yet commercial. pp. 51, 82, 98, 126–9, 137, 142 *et seq.*, 152, 161, 165, 171, 179, 181.

Bdellovibrio Stolp & Starr. Bacterium (Starr *et al.*, 1981). Parasitize other bacteria. Appear to be common and widespread in soil, but strains vary in pathogenicity. Difficult to culture. Not much studied and potential unknown. p. 88.

Cephalosporium Corda. Fungus, deuteromycete. A widespread genus, produce broad spectrum antibiotics. Not much studied. pp. 32, 123.

Cladosporium Link ex Fr. Fungus, deuteromycete (Ellis, M. B. *Dematiaceous hyphomycetes*, Kew: CMI). Common leaf surface saprophyte and occasionally pathogenic. Tried as an antagonist against some leaf diseases. pp. 68, 72–3, 81, 84–5, 98, 165, 167.

Colletotrichum Corda. Fungus, deuteromycete. *C. lagenarium, C. cucumarinum, C. lindemuthianum* are pathogens with host specificity (see pathogen index). A lot of studies in cross-protection, including induced resistance, but not commercial yet. pp. 32, 84–7.

Coniothyrium Corda. Fungus, deuteromycete (Sutton, B. C. (1980). *The Coelomycetes*, Kew: CMI; Ayers, W. A. & Adams, P. B. (1981). In *Biological control in crop production*, ed. G. C. Papavizas, pp. 91–105. Granada: Allenheld, Osmum Publishers). *C. minitans* parasitizes sclerotia. Widely distributed, easily grown to produce inocula and many studies made, but not commercial. p. 140.

Cryptococcus Kützing. Fungus, deuteromycete yeast (Lodder, J. (1970). *The yeasts*. Amsterdam: North Holland Publishing Co.). A common leaf surface fungus. Tried as an antagonist against various leaf diseases, used in field trials but not commercial. pp. 68, 72–3, 83.

Darluca Cast. Fungus, deuteromycete (perfect state = *Sphaerellopsis*). Parasitizes leaf surface fungi, especially rusts (Kranz, J. (1981). In *Microbial ecology of the phylloplane*, ed. J. P. Blakeman. London: Academic Press). Needs quite heavy infection of the pathogens, so reduces inoculum rather than helps prevent disease. pp. 24, 88–90.

Endothia Fr. Fungus, deuteromycete (see pathogen index). Vegetatively compatible hypovirulent strains of the pathogen are used to control chestnut blight. Large areas of forest now treated in Europe and also different strains now used in North America. pp. 102–4, 133.

Enterobacter Hormaeche & Edwards. Bacterium. *E. cloaca* is widely distributed in faeces of man and animals, soil and water. Tried as an antagonist, not commercial. pp. 129, 165, 179.

Erwinia Winslow *et al.* Bacterium (CMI 232; Starr *et al.*, 1981). *E. herbicola* is a common saprophyte on leaves and flowers. Experimentally used to control fire blight (see pathogen index, *E. amylovora*). Probably been superseded by *Pseudomonas* strains, especially ice⁻ ones. pp. 162–3.

Fusarium Link ex Fr. Fungus, deuteromycete (see pathogen index). *F. solani, F. oxysporum, F. lateritium and F. roseum* 'Gibbosum' are all used in various forms of cross-protection and competition, with avirulent strains or *formae speciales* on non-host cultivars. Especially worked on in some suppressive soils in France. Very common in soils worldwide. Also used as a wound protectant. pp. 26, 32, 82, 96–8, 106–7, 121–4, 126, 133, 167.

Gaeumannomyces Arx & Olivier. Fungus, ascomycete (see pathogen index). *G. graminis* var. *graminis* is avirulent on wheat and used for cross-protection together with related imperfect strains (*Phialophora, loc. cit.*). pp. 134–6.

Gliocladium Corda. Fungus, deuteromycete (Papavizas, G. C. (1985). *Annual Review of Phytopathology* **23**, 23–54). *G. virens* (perfect state *Hypocrea gelatinosa*, ascomycete) is a common soil saprophyte. Mycoparasite on a variety of fungi. Grows easily in culture and semi-commercial inoculum can be prepared. pp. 26, 105, 129, 171, 174, 177, 179.

Laccaria Berk. & Br. Fungus, basidiomycete (Harley, J. L. & Smith, S. E. (1983). *Mycorrhizal symbiosis*. London: Academic Press). *L. laccata* forms ectomycorrhizae on trees and may give protection from *Phytophthora cinnamomi*, not normally a biocontrol agent. p. 130.

Lactarius Pers. ex Gray. Fungus, basidiomycete. As *Laccaria* above. p. 130.

Leucopaxillus Boursier. Fungus, basidiomycete. As *Laccaria* above. p. 130.

Listeria Pirie. Bacterium (Starr *et al.*, 1981). Found in soil and on plants, mainly studied as a rare opportunistic pathogen. p. 80.

Microdochium Syd. Fungus, deuteromycete. *M. bolleyi* is being developed as a biocontrol agent against cereal root diseases (Kirk, J. J. & Deacon, J.W. (1987). *Plant and Soil* **98**, 231–7; Deacon, J. W. (1988). *Philosophical Transactions of the Royal Society* **B318**, 249–64). p. 158.

Myrothecium Tode ex Fr. Fungus, deuteromycete. Common saprotroph, used against damping-off. Not important. pp. 77, 123.

Penicillium Link ex Fr. Fungus, deuteromycete (CVS SM 23, see pathogen index). Very common saprotroph and occasional pathogen of fruits. Ruderal species, also competitive, produces antibiotics. Commonly occurs in lists of potential antagonists, especially from soil, but not commercially developed. pp. 82, 165, 167.

Peniophora Cooke. Fungus, basidiomycete. *P. giantea* (= *Phlebia gigantea*) was the first fungal biocontrol agent available commercially. Used against *Heterobasidion (Fomes) annosum* (Rishbeth, J. (1963). *Annals of Applied Biology* **52**, 63–77). Operates by hyphal interference and competitive exclusion of *Heterobasidion* from freshly cut stumps, especially on pine (*Pinus* spp.). pp. 43, 99–101.

Phialophora Medlar. Fungus, deuteromycete. Taxonomy of species confused (Walker, J. (1981). In *Biology and control of take-all* ed. M. J. C. Asher & P. J. Shipton, pp. 15–74. London: Academic Press). Usually imperfect states of *Gaeumannomyces* (see pathogen index and this index). *P. graminicola* and *P. hoffmanii* have been used to control take-all. Apparently work by competition for a similar niche on and in the root, by competitive exclusion and by induced resistance. pp. 39, 126, 134–6.

Phomopsis Sacc. Fungus, deuteromycete. Widespread genus as saprotrophs and plant pathogens. *P. oblongata* has been used in biological control of Dutch elm disease. p. 105.

Pisolithus Alb. & Schw. Fungus, basidiomycete. *P. tinctorus* is mycorrhizal, see under *Laccaria*. p. 130.

Pseudomonas Migula. Bacterium (Starr, *et al.*, 1981). The most widely used bacterial biocontrol agent, with much research done and in progress. Mostly ruderal species, good competitive ability, actively colonize roots, produce antibiotics and siderophores. About to be commercial for various root rots (e.g. 1988, *P. fluorescens* on cotton). *P. syringae* used for fire blight control and reducing ice damage (p. 90 *et seq.*). *P. fluorescens* and *P. putida* used against many other soil-borne diseases (e.g. take-all, p. 147). Also used against bacterial blotch of mushrooms (p. 49). Some strains are plant growth promoting (p. 151 *et seq.*), though probably by minor pathogen control rather than growth promotion *per se*. pp. 10, 16, 19, 32–3, 42, 49, 51, 68, 77–80, 82, 91–3, 105–6, 117–8, 125–9, 140, 142, 144 *et seq.*, 151, 153–5, 162–3, 166, 170–1, 175, 179–81.

Pythium Pringsh. Fungus, oomycete. Widely distributed, especially in wet soils. Some species cause damping off (see pathogen index), but *P. oligandrum* is a mycoparasite against other oomycetes and ascomycetes. Commercially available on a limited scale. pp. 123, 179.

Rhizobium Frank. Bacterium (Starr, *et al.*, 1981). Widely used for the formation of nitrogen fixing nodules (for which it is available commercially), but also gives some disease control, possibly by encouraging healthy growth with sufficient nitrogen. See general index. p. 1.

Sporidesmium Link ex Fr. Fungus, deuteromycete (CMI MP 70). *S. sclerotivorum* is a mycoparasite, especially against *Sclerotinia* spp. May be useful for reducing inoculum potential in the long term, rather than protection of a particular plant. pp. 18, 114, 140–1.

Sporobolomyces Kluyver & van Niel. Fungus, deuteromycete yeast (Lodder, J. (1970). *The yeasts*. Amsterdam: North Holland Publishing Co.). Common leaf surface saprotroph. *S. roseus* has been widely used on leaves to supplement natural populations and give disease control against a number of pathogens. Not commercial. pp. 68, 72–3, 77, 79, 83.

Streptomyces Waksman & Henrici. Bacterium (Starr, *et al.*, 1981). Widespread and common saprotrophs and some pathogens. *S. praecox* (= *Actinomyces praecox*) used to control potato scab. *S. griseus* also used against soil-borne diseases. Produce antibiotics. pp. 42, 82, 126–8, 143, 152, 171.

Suillus Karst. Fungus, basidiomycete. Forms mycorrhizae, see *Laccaria*. p. 130.

Thelephora Erhart ex Fr. Fungus, basidiomycete. Forms mycorrhizae, see *Laccaria*. p. 130.

Theratromyxa Protozoa, amoeba (Old, K. M. & Chakraborty, S. (1986). *Progress in Protistology* 1, 163–94). Mycophagous, shown to lyse *Cochliobolus*. p. 149.

Trichoderma Pers. ex Fr. Fungus, deuteromycete (CMI MP 116; Domsch, K. H., Gams, W. & Anderson, T-H. (1980). *Compendium of soil fungi*. Vol. 1. London: Academic Press; Papavizas, G. C. (1985). *Annual Review of Phytopathology* 23, 23–54; Chet, I. (1987). In *Innovative approaches to disease control*, ed. I. Chet. New York: John Wiley; Sivan, A. & Chet, I. (1986). In

Microbial communities in soil, eds. V. Jensen *et al.*, pp. 89–95. London: Elsevier). The most important and widely used fungal biocontrol agents are in this genus. Commercially available in various formulations and used to control many soil-borne diseases. All species are widespread and common soil saprotrophs, ruderal species and competitive, many produce antibiotics and they have a great range of enzymes to degrade polymers (including chitinases and cellulases). *T. viride* has been shown to control *Fusarium, Pythium, Rhizoctonia, Chondrosterium*, etc. (pp. 77, 138, 165, 173–4, 177). *T. harzianum* is a mycoparasite against *Fusarium, Rhizoctonia, Botrytis, Sclerotinia, Chondrosterium*, etc. (pp. 139, 172–4, 177–8). *T. hamatum* is also a mycoparasite (pp. 171, 173–7) and *T. koningii* and *T. pseudokoningii* are also used (pp. 77, 173). pp. 18, 24–5, 44, 82, 96, 105–6, 114, 123, 128–9, 137 *et seq.*, 157–8, 165–6, 170 *et seq.*, 173 *et seq.*, 181.

Tobacco mosaic virus Used in cross-protection against viruses and some fungi. p. 33.

Tubercularia Tode ex Fr. Fungus, deuteromycete. *T. vinosa* has been tried as an antagonist, not important. p. 89.

Tuberculina Sacc. Fungus, deuteromycete. Used for control of rusts. p. 24.

Ulocladium Massel, Fungus, deuteromycete. Has been tested for biocontrol on leaf surfaces, not important. p. 167.

Vampyrella Cienk. Protozoa, amoeba. See references under *Acanthamoeba*. Parasitizes fungi, importance unknown. pp. 24, 149.

Verticillium Nees ex Wallr. Fungus, deuteromycete. Common and widespread in soil. Some species are pathogenic (see pathogen index), others saprotrophs (Domsch, K. H., Gams, W. & Anderson, T-H. (1980). *Compendium of soil fungi*. Vol. 1. London: Academic Press). *V. albo-atrum* (avirulent strain) and *V. dahliae* (avirulent strain) have both been used for cross-protection studies (p. 84). *V. lecanii* was originally used against insects (and is available commercially for this) but it is also a mycoparasite on some rusts. pp. 24, 89, 133–5.

Xanthomonas Dowson. Bacterium (Starr, *et al.*, 1981). A common leaf surface saprotroph, has been tried for biological control, not important. pp. 68, 80.

Subject index

Refer to the expanded index of pathogens (p. 200) and the expanded index of antagonists (p. 207) for reference to most micro-organisms. This subject index only includes those microbes whose importance to plant disease is peripheral.

dispersal, 9, 28
dormant survival, 20
dithiocarbamates, 70
drench, defined, 186
dsRNA, 102, 105, 173
  defined, 186
Dutch elm disease, 95, 104

ectomycorrhizae, 130
endaphic, 1, 186
EDDHA, 145
EDTA, 145
elicitor, 86
  defined, 186
elm, 104
emergence promoting bacteria, 154
endolysis, 20 *et seq.*
Enterobacter, 155
environment, defined, 186
environmental control, 30
environmental effects of pesticides, 45
environmental protection, 56
environmental stress, leaves, 77
epidemiology, 8
  defined, 186
epinasty, 155
Escherichia, 93
ethene, 21, 155
ethirimol, 70
eucalyptus, 130
exolysis, 20, 24 *et seq.*, 28
extreme environments, 2
exudates from seeds, 171
  from roots, 113
eyespot of cereals, 35, 95, 157

farmyard manure, 37
fermenters, 57
field testing biocontrol agents, 59
field trial, 137
fire blight, 91, 95, 162, 163, 167
Flavobacterium, 16, 155
flax, 69, 144
flowers, 161 *et seq.*
fluorescent pseudomonad, *see*
  Pseudomonas in pathogen and
  antagonist index
food webs, 2
foot rot of cereals, 157
forma specialis, defined, 186
formulation, 138, 152
  defined, 186
  of control agents, 62
  of inoculum, 170, 173, 176, 182
frost damage, 90, 93
fruit, 161 *et seq.*
  trees, 106
  tree diseases, 157
fumigant, defined, 186
fumigation, 157

fungicide, defined, 186
  wide spectrum, 71
fungicide resistance, 46, 64, 71, 158, 165
fungicide selectivity, 70 *et seq.*
fungicides on non-target organisms, 70 *et seq.*
fungistasis, 19, 26, 78, 121, 127
  defined, 186
Fusarium suppressive soil, 120 *et seq.*

general antagonism, defined, 186
general suppression, defined, 186
general suppressiveness, 127
genetic engineering, 8, 44, 55, 92–3, 154, 181
  defined, 187
genetics, host plant, 2
germination, 28
germination inhibition, 10, 78
germination lysis, defined, 187
germination of spores, 27
germination promoting bacteria, 154
glasshouse crops, 49
Glomus, 132
gnotobiotic, defined, 187
gnotobiotic plants, 58
grazing by arthropods, 175
  by collembola, 175
  by protozoa, 179–80
green manure, 7, 26, 28, 37, 41–2, 50
  defined, 187
Grossglockneria, 24
groundnut, 139
growth hormones and growth promotion,
  58, 151
gums, 17

habitat, defined, 187
harmful micro-organisms, 154, *see*
  deleterious bacteria
honey fungus, 157
horticultural crops, 49, 159
host cultivar effect, 183
host defence, 114
host defence mechanisms, 29 *et seq.*
host defence reactions, 27
host immune, 30
host metabolism, 30
host recognition systems, 27, 31
host resistance, 29, 30, 39, 94
host susceptible, 30
hydrogen cyanide, 21
hyperparasites, *see* mycoparasites
hyperparasitism, defined, 187
hypersensitive response, 31
hyphal interference, 100–1
hypovirulence, 102–3, 172 *et seq.*
  defined, 187

iatrogenic disease, defined, 187
ice damage, 162